CAMBRIDGE MONOGRAPHS ON MATHEMATICAL PHYSICS

General editors: P. V. Landshoff, W. H. McCrea, D. W. Sciama, S. Weinberg

GROUP STRUCTURE OF
GAUGE THEORIES

GROUP STRUCTURE OF GAUGE THEORIES

L. O'RAIFEARTAIGH

Dublin Institute for Advanced Studies

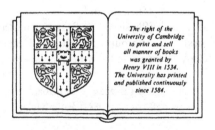

The right of the
University of Cambridge
to print and sell
all manner of books
was granted by
Henry VIII in 1534.
The University has printed
and published continuously
since 1584.

CAMBRIDGE UNIVERSITY PRESS

CAMBRIDGE

NEW YORK   NEW ROCHELLE

MELBOURNE   SYDNEY

CAMBRIDGE UNIVERSITY PRESS
Cambridge, New York, Melbourne, Madrid, Cape Town, Singapore,
São Paulo, Delhi, Dubai, Tokyo, Mexico City

Cambridge University Press
The Edinburgh Building, Cambridge CB2 8RU, UK

Published in the United States of America by
Cambridge University Press, New York

www.cambridge.org
Information on this title: www.cambridge.org/9780521347853

First published 1986
First paperback edition (with corrections) 1988

A catalogue record for this publication is available from the British Library

Library of Congress Cataloguing in Publication Data

O'Raifeartaigh, L. (Lochlainn)
Group structure of gauge theories
(Cambridge monographs on mathematical physics)
Bibliography
Includes index
1. Gauge fields (Physics). 2. Broken symmetry
(Physics). 3. Lie groups. I. Title. II. Series.
QC793.3.F5069 1985 530.1′43 85–7828

ISBN 978-0-521-25293-5 Hardback
ISBN 978-0-521-34785-3 Paperback

Contents

Preface

It has been known for many years that the gravitational and electromagnetic interactions of matter can be formulated as gauge theories – based on the Lorentz group $SO(3, 1)$ and the compact 'internal' phase group $U(1)$, respectively. But over the past two decades it has gradually come to be accepted that the remaining two (known) fundamental interactions of matter, namely the strong and weak nuclear interactions, are also gauge interactions, a property that had been hidden by confinement for the strong interactions and by spontaneous symmetry breaking for the weak ones. To be more precise, it has now been established beyond reasonable doubt that the weak nuclear interactions combine with electromagnetism to form a gauge interaction based on the compact internal non-abelian group $U(2)$, and, although the evidence is less direct, it is accepted that the strong interactions are gauge interactions based on the compact simple internal (colour) group $SU(3)$. The upshot of these results is that the (known) non-gravitational interactions are now described by a gauge theory based on a compact internal group with Lie algebra $SU(3) \times SU(2) \times U(1)$. (The global group is actually $S(U(3) \times U(2))$ because of certain discrete correlations in the particle classification, chapter 9.)

If the $S(U(3) \times U(2))$ theory of the non-gravitational interactions is correct, it represents an immense advance because gauge theories, by their nature, determine the form of the interactions, leaving only a finite number of constants as free parameters. In fact it means that the form of all the fundamental interactions is now known. Furthermore, since gauge theories have a geometrical interpretation in terms of fibre bundles, it means that even the non-gravitational interactions have a geometrical significance and are thus brought a step nearer to gravitation.

On the other hand the fact that all the interactions have a common gauge structure does not mean that they are fully unified, because the gravitational interaction has special properties not shared by the others (the existence of the metric and the equivalence principle, for example) and the three other interactions remain separate in the sense that the $SU(3) \times SU(2) \times U(1)$ algebra consists of three irreducible pieces, with a

vii

separate coupling constant for each piece. For this reason it has been suggested that the gauge group $S(U(3) \times U(2))$ is actually only a subgroup of a larger, simple, compact gauge group G, which has only one coupling constant and which truly unifies the three non-gravitational interactions. Theories based on such groups G are called grand unified theories (GUTs) and have been extensively studied in recent years. Although the most spectacular prediction of GUTs, namely proton decay, has not yet been (and may even never be) observed, there is a certain amount of indirect evidence for GUTs (chapter 10), notably from the particle classification, from renormalization group considerations and from cosmology.

Compact gauge theories are, in principle, generalizations of electro-magnetism from $U(1)$ to non-abelian groups, but the generalization is not trivial for two reasons. First, the intrinsic group structure (Lie algebras, representations, invariants, etc.) is much more complicated than in the abelian case. Second, spontaneous symmetry breaking, which enters only in the special case of superconductivity for electromagnetism, plays a central role for the non-abelian theories.

The aim of the present monograph is to provide a review of the group structure both of the non-abelian gauge theories themselves and of their spontaneous symmetry breaking. The presentation is pitched at about the graduate student level and, so as not to overlap with the many excellent treatments of other aspects of gauge theories (renormalization, phenomen-ology, confinement, topology, etc.), it concentrates on two aspects. These are the group theoretical background, particularly the global group theory (Part I) and the algebraic structure of the gauge interactions and of their symmetry breakdown patterns (Part II). The spontaneous symmetry breaking is treated in some detail (at the classical level) because many results in this area have not previously been available in book form. It should be stated, however, that the investigation of symmetry breaking patterns is still at an early stage of development and so the results presented should be regarded as pioneering ones.

The general plan of the monograph may be seen from the list of contents, but a few remarks may be in order. In chapters 1–5, where the group-theoretical background is given, some of the more technical equipment (tables of branching rules and Clebsch–Gordon coefficients for example) has been omitted because it is available elsewhere and space did not permit a reasonable resumé. In the chapters on spontaneous symmetry breaking (8, 11, 12) it is assumed, for definiteness, that the symmetry breakdown is caused by a local scalar potential, but it is fairly evident that because of the group-theoretical nature of the results most of them would survive in a much broader context, e.g. if the scalar field were composite. Indeed this is one of the justifications for the group-theoretical approach. With regard to the references, and the suggestions for further reading, the literature on both Lie groups and gauge theory is so vast there was no hope of providing a comprehensive bibliography, and accordingly these sections have been limited to those references which are strictly relevant, to recent reviews (many of which, notably Langacker (1981), contain further lists of references) and to books.

Finally, I should like to take this opportunity to thank Professors Nikolas Kuiper (director) and Louis Michel for their kind hospitality at the Institut des Hautes Etudes Scientifiques, Bures-sur-Yvette, for most of the academic year 1983–4, when much of the monograph was written. I should also thank Louis Michel, whose influence pervades not only the book but the whole literature on symmetry and symmetry breaking, for many invaluable discussions and comments.

L. O'RAIFEARTAIGH   *Dublin, 1985*

Part I
Group structure

1
Global properties of groups and Lie groups

1.1 Groups

Part I of this monograph is concerned mainly with the theory of Lie groups, which is the background group theory necessary for gauge interactions. But since some of the relevant properties of Lie groups are common to many topological groups, and even to groups in general, it is convenient to begin with a brief discussion of general and topological groups.

First, a group G is defined as a set of elements $\{g\}$ (not necessarily finite or even countable) for which there is a multiplication law $G \times G \to G$ ($g_1 g_2 \in G$ for $g_1, g_2 \in G$) with the three properties:

(i) Associativity, $g_1(g_2 g_3) = (g_1 g_2)g_3$;
(ii) Existence of an identity element e, such that, for each $g \in G$, $eg = ge = g$;
(iii) Existence of an inverse element g^{-1} such that, for each $g \in G$, $g^{-1}g = gg^{-1} = e$.

The identity e and the inverse g^{-1} are easily seen to be unique. In physics and geometry, groups usually occur as transformations which leave some quantity (or set of quantities) invariant, simply because the product of any two such transformations also leaves the quantity invariant. In particular, groups of transformations that leave the Hamiltonian or Lagrangian invariant are called symmetry groups.

An important concept which arises for group transformations is that of group *orbits*, which may be defined as follows: let a group G of transformations act on a set of elements $\{s\}$. Then the subset of all elements that can be obtained from any given element s_0 by the action of G is called the *orbit* of s_0 with respect to G, or the G-orbit of s_0. For example, the orbits of the rotation group in Euclidean 3-space are the surfaces of the spheres of radius r for each $0 \leqslant r < \infty$. The group itself is an orbit with respect to left or right multiplication, and group invariants are trivial (one-element) orbits. The action of the group on an orbit is transitive (i.e. any element on an orbit can be obtained from any other one by a group transformation)

3

and membership of an orbit is a class relation ($s \in O(s_0) \Rightarrow s_0 \in O(s)$ and $s \in O(s'), s' \in O(s'') \Rightarrow s \in O(s'')$). Thus any set on which a group acts can be partitioned into distinct group orbits.

The set of elements z that commute with all other group elements ($zg = gz$ all $g \in G$) is called the *centre* Z of G, and it includes at least the identity e. If all elements commute ($Z = G$) the group is called *abelian*.

A *subgroup* H of a group G is a subset of $\{h\}$ of elements of G which closes with respect to the multiplication already defined by G ($hk \in H$ for $h, k \in H$) and which contains the inverse of each of its elements h, and the identity e. Subgroups usually arise by making a natural restriction of the original group, e.g. restricting the group of rotations in three dimensions to rotations about one axis, or to discrete rotations.

One of the most important group operations is *conjugation*. The conjugation of one element g by another h is the transformation $g \rightarrow hgh^{-1}$. The conjugation of the group G by a single element h is the transformation $g \rightarrow hgh^{-1}$ for all $g \in G$ and it preserves the group multiplication since $gg' \rightarrow h(gg')h^{-1} = (hgh^{-1})(hg'h^{-1})$. The set of elements obtained by the conjugation of a fixed element h by the whole of G, ghg^{-1} for all $g \in G$, is called the conjugation *class* of h, and, since the class is a group orbit, a group can be partitioned into distinct conjugacy classes. For example, for the permutation group of three objects (in standard notation), the conjugacy classes are $\{(1), (2), (3)\}$ $\{(12), (13), (23)\}$ and $\{(123), (132)\}$. Since $geg^{-1} = e$, the identity element forms a separate conjugacy class, and more generally, a given element forms a separate conjugacy class if, and only if, it lies in the centre Z. In physical or geometrical situations the elements of a conjugacy class usually have some obvious physical or geometrical characteristic in common. For example, for the rotation group all rotations of the same magnitude (but in different directions) are in the same conjugacy class.

An invariant subgroup H of G is one which is invariant with respect to conjugation with G, i.e. $ghg^{-1} \in H$ for all $h \in H$, $g \in G$. For example, the translation subgroups of the space–time groups are invariant subgroups because they are transformed into themselves by rotations.

To each subgroup H of a group G is associated its right (left) cosets, where the cosets are the G-orbits of H with respect to right (left) multiplication. In other words, g_1 and g_2 are in the same right (left) coset if, and only if, there exists an element h of H such that $g_1 = g_2 h$ (hg_2). From the properties of orbits it follows that the group may be partitioned into distinct cosets, and because $g_1 = hg_2 \Leftrightarrow h = g_1 g_2^{-1}$ and $g_1 g_2^{-1} = e \Leftrightarrow g_1 = g_2$ the dimension of each coset, including H itself, is the same. For finite groups the equidimensionality of the cosets implies that dim G/dim H = integer (= number of cosets) and thus gives a limitation on the number of subgroups. In geometrical or physical situations the cosets are often parametrized in a direct geometrical or physical manner. For example, for fixed mass, the cosets of the rotation subgroup $SO(3)$ of the Lorentz group $SO(3, 1)$ are often parametrized by the 3-momentum **p**.

In general the left and right cosets of a subgroup H are not identical. If they are, and if h and g are any elements of H and G respectively, then $gh = h'g$ for some $h' \in H$ and thus $ghg^{-1} = h'$, which is just the condition that H be an invariant subgroup. Conversely, if $ghg^{-1} = h' \in H$ for all $g \in G$, $h \in H$, then $gh = h'g$. Thus the necessary and sufficient condition for the left and right cosets of a subgroup to be the same is that the subgroup be invariant.

When H is an invariant subgroup the cosets c themselves form a group, called the quotient group $Q = G/H$. For the quotient group the identity element is H itself and the multiplication is that induced by G, i.e. $cc' = c''$ where $g \in c$, $g' \in c'$ and c'' contains gg'. Note that invariance of H is necessary for the consistency of this scheme. The quotient group is not a subgroup of G, nor is it, in general, isomorphic to a subgroup of G. An interesting example from physics is the Heisenberg group of real upper triangular 3×3 matrices:

$$H(\alpha, \beta, m) = \begin{pmatrix} 1 & \alpha & m \\ 0 & 1 & \beta \\ 0 & 0 & 1 \end{pmatrix}. \tag{1.1}$$

The subgroup $\alpha = \beta = 0$ is an invariant (even central) subgroup, and the quotient group is the two-dimensional translation group

$$T(\alpha, \beta) \approx \begin{pmatrix} 1 & \alpha \\ 0 & 1 \end{pmatrix} \times \begin{pmatrix} 1 & \beta \\ 0 & 1 \end{pmatrix}.$$

But the quotient group is not a subgroup of $H(\alpha, \beta, m)$ and both the invariant subgroup and the quotient group are abelian although $H(\alpha, \beta, m)$ itself is not.

A mapping of a group onto itself $(g \rightarrow g'(g) \in G)$ which preserves the multiplication law is called an *automorphism* of the group. For example, for the group of unitary unimodular complex $n \times n$ matrices $SU(n)$, $n \geqslant 2$, complex conjugation is an automorphism, but hermitian conjugation is not because it reverses the order of multiplication. It has been seen already that the conjugation of a group by a fixed element preserves the multiplication law. Thus conjugation with any element is an automorphism, and any automorphism that can be implemented by a conjugation is called an *inner automorphism*. Thus complex conjugation $M \to M^*$ is inner for $SU(2)$ since $M^* = CMC^{-1}$ where $M \in SU(2)$ and $C = \begin{pmatrix} 0 & 1 \\ -1 & 0 \end{pmatrix} \in SU(2)$, but is not inner for $SU(n)$, $n \geqslant 3$ because for $n \geqslant 3$ there are elements such that $\operatorname{tr} M \neq \operatorname{tr} M^*$ and for such elements M and M^* cannot be conjugate. The sets of all automorphisms $(\operatorname{Aut}(G))$ and of all inner automorphisms $(\operatorname{Int}(G))$ of a group G are themselves groups.

A map from a group G onto (or indeed into) another group G' such that the multiplication law is preserved is called a homomorphism of G, and G' is called the homomorphic image. The set of elements of G which map onto the identity element e' of G' is called the *kernel* of the homomorphism. For example, the rotation group $SO(3)$ in three dimensions is a homomorphic image of the Euclidean group $E(3)$ (rotations and translations) with the translation subgroup of $E(3)$ as the kernel. It is not difficult to see that the kernel of a homomorphism $G \to G'$ must be an invariant subgroup of G. If the homomorphism between G and G' is one–one the two groups are said to be *isomorphic*.

A *direct product* $G = A \times B$ of two groups A and B is a group whose elements are $g = (a, b)$ for all $a \in A$, $b \in B$ and whose multiplication law is $gg' = (aa', bb')$. A somewhat more subtle structure is that of a *semi-direct* product. A fairly familiar occurrence of this structure is in the case of the Euclidean group $E(3)$ and its various crystallographic (finite) subgroups, all of which are semi-direct products of rotation groups R and translation groups T. That means that each element g of the group can be expressed as $g = (r, t)$ where r is a rotation and t is a translation and the multiplication law is $gg' = (r, t)(r', t') = (rr', t+rt')$ where rt' denotes the translation t' rotated by r. Note that the rotations affect the translations but not vice versa. The generalization of this law for a semi-direct product $A \wedge B$ of any two groups A and B is $g = (a, b)$ where a and b are elements of A and B and $gg' = (a, b)(a', b') = (aa', bh(a)b')$ where the transformations $b \to h(a)b$ are automorphisms of B, i.e. $(h(a)b)(h(a)b') = h(a)bb'$ and are homomorphisms (representations) of A, i.e. $h(aa')b = h(a)(h(a')b)$. This multiplication law becomes more intuitive in the special case that the automorphisms of B are inner, each $a(A)$ has an image $b(a)$ in B and $(a, b)(a', b') = (aa', bb(a)b'b(a)^{-1})$, in particular the fact that the multiplication law satisfies the associativity condition can be seen by inspection. The subgroup $\tilde{B} = (1, B)$ of a semi-direct product $A \wedge B$ is evidently an invariant subgroup and the subgroup $\tilde{A} = (A, 1)$ is a group of representatives of the cosets. Conversely, if a group G has an invariant subgroup H and there exists a set of representatives for the cosets G/H that form a group K, then G is a semi-direct product of the form $K \wedge H$. A relatively simple illustration is provided by the groups $O(2)$ and $N(2)$, where $N(2)$

is the maximal subgroup of $SU(2)$ that conjugates $SO(2) \subset SU(2)$ into itself. One sees that both groups have $SO(2)$ as an invariant subgroup and have just one other coset (with representative R, say). But since $R^2 = I$ and $R^2 = -I$ for $O(2)$ and $N(2)$ respectively, (I, R) is a group only for $O(2)$ and thus $O(2)$ is a semi-direct product while $N(2)$ is not.

1.2 Topological groups

When the elements of a group are not denumerable it is often desirable to have a notion of continuity, and so a topology is introduced. A topological group is defined to be a group with any topology in which the multiplication and inversion are continuous. That is to say, if g and h are in the neighbourhood of g' and h' respectively, in the given topology, then gh and g^{-1} are in the neighbourhood of $g'h'$ and $(g')^{-1}$. In practice, a topology is often suggested by the context in which the group is considered. Thus, for example, a natural topology for a group of real matrices is the usual topology of the real line for the matrix elements.

The most important topological concept that will be needed is that of *compactness*. It will be recalled that a topological space is compact if every covering (set of open sets containing every point) has a finite subcovering. In particular, for a space whose topology is induced by a metric, compactness is equivalent to the statement that the space is closed and bounded with respect to that metric. A much weaker, but very important form of compactness is *local* compactness, and a topological space is said to be locally compact if every neighbourhood of a point contains a compact subneighbourhood. Lie groups, which are locally Euclidean, are locally compact but not necessarily compact.

One of the first uses of a topology is to define the connectivity structure of a group. Two elements g and h are said to be connected, if for a real parameter $0 \leqslant t \leqslant 1$ there exists in G a continuous path $g(t)$ with $g(0) = g$ and $g(1) = h$. The connectedness of elements is a class relation and hence a topological group may be partitioned into disjoint self-connected components. The component G_0 containing the identity element e is called the identity component. For example, the group $O(n)$ of real orthogonal $n \times n$ matrices, $O^t(n)\, O(n) = 1$, where t denotes transpose, consists of two disconnected components $\det O(n) = 1$ and $\det O(n) = -1$, the first, called $SO(n)$, being the identity component.

In the $O(n)$ example one sees that the identity component is an invariant subgroup ($\det N = 1$ implies $\det NK = 1$ for $\det K = 1$ and $\det MNM^{-1} = 1$ for $\det M = \pm 1$), and it turns out that this result is completely general: *the identity component of any topological group is an invariant subgroup.* The proof is quite straightforward. First, the connected component forms a subgroup because, if g and h are connected to e (by $g(t)$ and $h(t)$), then gh is connected to e (by $g(t) h(t)$). Second, the subgroup is invariant because, for g connected to e and any $k \in G$, kgk^{-1} is connected to e (by $kg(t) k^{-1}$).

The other components are the cosets of G_0 because for g, h in the same component (connected to each other by $k(t)$ say) $g^{-1}h$ is in the connected component (connected to e by $g^{-1}k(t)$). Thus the components form a discrete quotient group $D = G/G_0$. In many cases the group G is actually a direct or semi-direct product of the form $D \times G_0$ or $D \wedge G_0$. For example for the orthogonal groups, $O(2n+1)$ is a direct product of the form $Z_2 \times SO(2n+1)$ where Z_2 is the two-element group ± 1, and $O(2n)$ is a semi-direct product of the form $Z_2 \wedge SO(2n)$ where Z_2 is the two-element group $\mathrm{diag}(1, 1, ..., 1, 1, \pm 1)$. However, G is not always a direct or semi-direct product, even for compact groups. For compact Lie groups a complete analysis of the disconnected structure has been given by de Siebentahl (1956).

A connected group is said to be *simply connected* if each closed continuous curve $g(t)$ ($0 \leqslant t \leqslant 1$, $g(0) = g(1)$) in it may be continuously deformed to zero (by a family of such curves $g(t, s)$, $0 \leqslant s \leqslant 1$ $g(t, 0) = g(t)$, $g(t, 1) = e$). (For those familiar with homotopy theory (Nash and Sen, 1983) simple connectivity $\Leftrightarrow \pi_1(G) = 0$.) For example the $1 + 1$ dimensional Lorentz group e^x, $-\infty < x < \infty$ is simply connected, but the rotation group $e^{i\phi}$, $0 \leqslant \phi \leqslant 2\pi$ is not. For Lie groups, simple connectivity will be discussed in more detail in section 1.4. Another important use of a topology is to construct a measure $\mu(g)$ on the group. A measure is desirable because it allows the concept of summing over the elements for finite groups to be generalized to integration for continuous groups. For this reason the measure is required to be invariant with respect to group multiplication (either left or right). That is, it is required to satisfy the relation

$$\int d\mu(g) f(gh) = \int d\mu(g) f(g) \qquad (1.2)$$

(and similarly for left multiplication) for every continuous function $f(g)$ of compact support. A sufficient condition for the existence of such a measure is that the group be locally compact (Weil, 1953), in which case the measure is called Haar measure and is unique up to a constant. For abelian groups, the left and right Haar measures are the same, and for the Euclidean translation groups it is just the Lebesgue measure. Similarly, for the Heisenberg group (1.1) both measures are just the Euclidean measure $d\alpha \, d\beta \, dm$, and for the general real linear group $GL(n, r)$ both measures are

$(\det M)^{-1} \prod dm_{ij}$, where m_{ij} are the matrix elements of M. However, for the real triangular matrix group $\begin{pmatrix} e^a & x \\ 0 & e^b \end{pmatrix}$, the left and right measures are different, being $\exp(-a)\,da\,db\,dx$ and $\exp(-b)\,da\,db\,dx$ respectively. Since Lie groups are locally compact they always have invariant measures, and an explicit construction will be given in section 2.2.

In the non-abelian case a simple criterion for the left and right Haar measures to be the same is the following: let $\mu(g)$ be a left-invariant measure. Then, for any fixed element k, $\mu(gk)$ is also a left-invariant measure, and hence, by the uniqueness, $\mu(gk) = \lambda(k)\mu(g)$ where $\lambda(k)$ is a positive factor, called the modular factor. Then

$$\int d\mu(g) f(gk^{-1}) = \int d\mu(gk) f(g) = \lambda(k) \int d\mu(g) f(g), \qquad (1.3)$$

and so $\mu(g)$ is right-invariant if, and only if, the modular factor is a constant.

Compact groups have the special property that the continuous functions of compact support include $\theta(g) \equiv 1$. Using θ for f in (1.3) one sees at once that $\lambda(k) = 1$ and thus for compact groups the left and right measures are the same. Furthermore, since

$$\int_G d\mu(g) = \int d\mu(g)\,\theta(g) < \infty, \qquad (1.4)$$

the total measure for compact groups is finite. This is probably the most important property of compact groups and one of its consequences is that the representation theory of compact groups is similar to that of finite groups (chapter 5).

It is often useful to have a left- (right-)invariant *metric* $\rho(g, h) = \rho(kg, kh)$ and a necessary and sufficient condition for this is that the topology have a countable basis. It is necessary because any metrically induced topology has such a basis (e.g. the spheres $\rho(g, g) <$ rationals) and is sufficient because of a theorem due to Birkhoff and Takhutani (see Barut and Raczka 1977). Since Lie groups are locally Euclidean (section 1.3) they satisfy this condition, and the metric will be constructed explicitly in section 2.2.

For compact groups, in particular compact Lie groups, there exist metrics which are both left- and right-invariant. In fact, for compact groups any measurable metric can be converted into a left- or right-invariant one by averaging with the Haar measure. Thus, in particular, if $\rho(f, g)$ is a left-invariant metric,

$$\Delta(f, g) = \int d\mu(h)\,\rho(fh, gh), \qquad (1.5)$$

is again a metric, and is both left- and right-invariant.

Finally it should be mentioned that by subgroups of topological, in particular Lie, groups will be meant closed subgroups, where closed means closed in the topology of the original group. Thus, for example, the subgroups obtained by restricting the continuous parameters of Lie groups to the rationals are excluded.

1.3 Lie groups: global considerations

After the above digression into general and topological groups let us consider Lie groups. These are groups for which the topology is locally Euclidean. That is, in the neighbourhood of any point the group may be parametrized by a *finite* number of *continuous* (real) variables. Thus $g = g(a_1, \dots, a_r)$, where the a_k, $k = 1, \dots, r$ are continuous and, by convention, $e = g(0, 0, \dots, 0)$. In general the 'rigid' groups of theoretical physics, e.g. the rotation groups, Lorentz group, groups of unitary transformations on finite-dimensional spaces etc. are Lie groups, while the more 'flexible' groups such as the group of all coordinate transformations in general relativity, local gauge groups, and the group of all canonical transformations in classical point mechanics are not Lie groups because the number of parameters is not finite. The archetypal Lie group is an $n \times n$ matrix group, with continuous elements, and indeed it can be shown that every Lie group is isomorphic to a matrix group (at least in the neighbourhood of the identity). Thus it is useful to keep the continuous matrix groups in mind as concrete realizations of Lie groups. Some important examples of continuous matrix groups which are Lie groups are:

(1) $GL(n, c/r)$ = group of all complex/real non-singular $n \times n$ matrices M.

(2) $SL(n, c/r)$ = group of all complex/real unimodular $n \times n$ matrices M (det $M = 1$).

(3) $D(n, c/r/u)$ = group of all complex/real/unitary diagonal non-singular $n \times n$ matrices.

(4) $T(n, c/r)$ = group of all complex/real upper-triangular non-singular $n \times n$ matrices.

(5) $T_0(n, c/r)$ = group of all complex/real upper-triangular unit-diagonal $n \times n$ matrices.

(6) $U(n)$ = group of all complex unitary $n \times n$ matrices.

(7) $SU(n)$ = group of all complex unitary unimodular $n \times n$ matrices.

(8) $O(n, c/r)$ = group of all complex/real orthogonal $n \times n$ matrices.

(9) $SO(n, c/r)$ = group of all unimodular complex/real orthogonal $n \times n$ matrices.

(10) $Sp(2n, c/r/u)$ = group of all complex/real/unitary symplectic $2n \times 2n$ matrices M, i.e. $M^t J M = J$ where t = transpose, $J = \begin{pmatrix} 0 & 1 \\ -1 & 0 \end{pmatrix}$ and 1 is the unit $n \times n$ matrix.

(11) $SU(p, q)$ = group of all pseudo-unitary, unimodular $n \times n$ matrices M, i.e. $M^\dagger G M = G$ where $G = \begin{pmatrix} 1_p & 0 \\ 0 & -1_q \end{pmatrix}$ and 1_p and 1_q are the p- and q-dimensional unit matrices.

(12) $O(p, q)$ = group of all pseudo-orthogonal $n \times n$ matrices M, $M^t G M = G$.

Note that the various conditions that have been placed on the general matrix group $GL(n, c)$ to obtain the other groups are compatible with the multiplication laws. Thus the product (or inverse) of two real/unitary/upper-triangular etc. matrices is again real/unitary/upper-triangular, and so on. Note also that (real, unitary) = (real, orthogonal). The unitary unimodular groups $SU(n)$ and its subgroups $SO(n, r)$ and $Sp(2n, u)$ are called the *classical* Lie groups and it is perhaps interesting to note that they may be thought of as the groups of unitary matrices with complex, real and quaternionic entries respectively.

The minimum number of parameters necessary to parametrize a Lie group is called the *order* r of the group. Thus the order of $GL(n, c)$ is $r = 2n^2$; that of $SL(n, r)$ is $r = n^2 - 1$; that of $SO(n, r)$ is $r = \frac{1}{2}n(n-1)$ and so on. The r-dimensional parameter space may be compact or non-compact, connected or disconnected and the topology of the Lie group is just the topology of this space. In the matrix realization the only compact Lie groups are those whose elements can be represented by unitary matrices, i.e. $U(n)$ and its subgroups.

The essential features of Lie groups, however, are the differential, or local, features, and in order to separate these from global properties such as connectedness, the concept of *local isomorphism* is introduced. Two Lie groups are said to be locally isomorphic if they are isomorphic in a neighbourhood of the origin in parameter space. The point of this concept is that it can be shown (Pontryagin, 1966) that in each set of locally isomorphic connected Lie groups G there is a unique one $\tilde{G}$ which is simply connected, and that all the others are homomorphic images of this one. The simply connected member $\tilde{G}$ of the set is called the *covering* group for the set because its parameter space covers that of the other members of the set an integer number of times (like Riemann sheets for complex variables). For example, for the two-dimensional rotation group $R(\phi)$ where $0 \leqslant \phi \leqslant 2\pi$ the covering group $\tilde{G}$ is the group of real numbers x with respect to addition, the homomorphism being given by the map $\phi = x \bmod 2\pi$, and for the connected three-dimensional rotation group

$SO(3)$ the covering group is $\tilde{G} = SU(2)$, the homomorphism being given by the map

$$u \to R_{ij} = \text{tr}(u^\dagger \sigma_i u \sigma_j), \quad u \in SU(2), R \in SO(3), \qquad (1.6)$$

where σ_i, $i = 1, 2, 3$ are the Pauli matrices.

In the above examples the kernel of the homomorphism (the set of elements in the covering group $\tilde{G}$ which map into the identity of the image) is a discrete group lying in the centre of $\tilde{G}$ (i.e. commuting with all the elements of $\tilde{G}$), and it is not difficult to see that this is a general result: The kernel of the homomorphism $\tilde{G} \to G$ is always discrete and central. In fact if the kernel K were not discrete, then $\tilde{G}$ would have more continuous parameters than G, and if it were not central then (since $\tilde{G}$ is connected) some elements $k \in K$ would not commute with elements in the neighbourhood of the identity, and so could be changed continuously by conjugation with these elements, which contradicts the discreteness. Thus, in effect, any multiply connected Lie group may be written in the form $\tilde{G}/Z$ where $\tilde{G}$ is simply connected and Z is a discrete abelian group in the centre of $\tilde{G}$.

By combining this result on simple connectivity with the result of the previous section on connectivity one sees that every Lie group has the following global structure:

(1) It consists of a discrete number of disconnected components, the identity component G_0 (which of course is connected) being an invariant subgroup and the other components its cosets.

(2) The quotient group $D = G/G_0$ is discrete, and although G is not necessarily a 'semi-direct product of the form $G = D \wedge G_0$ it often is (especially when G is compact).

(3) The identity component G_0 is connected and is the homomorphic image of a simply connected (covering) group $\tilde{G}_0$.

(4) The kernel of the homomorphism is a discrete group Z in the centre of $\tilde{G}_0$, so that actually $G_0 = \tilde{G}_0/Z$, where Z is a discrete central group.

Exercises

1.1. Show that the kernel of a homomorphism $G \to G'$ must be an invariant subgroup of G.

1.2. Show that the multiplication law $gg' = (a, b)(a, b') = (aa', bh(a)b')$ for a semi-direct product satisfies the associativity condition $g(g'g'') = (gg')g''$.

1.3. Show that $SU(2)$ is simply connected. (Hint: show that it is topologically equivalent to S_3, the surface of the unit sphere in four dimensions.)

1.4. Show that $SO(2)$ is abelian, but $O(2) = Z_2 \wedge SO(2)$ is not, and find its conjugacy classes.

2
Local properties of Lie groups

2.1 Structure functions and structure constants

Let us now consider the local structure of Lie groups. The local structure contains the most essential feature of the groups, namely their continuity, and it is studied by considering the group for values of the parameters in the neighbourhood of the identity. The remarkable property of Lie groups, first discovered by Lie himself, is that in spite of the continuity of the parameters, the study of the local structure can be reduced to the study of finite discrete systems, namely Lie algebras. For example, the local structure of the rotation group $SO(3)$, for which the Euler angles ϕ, θ, ψ are the continuous parameters, is completely determined by the well-known Lie algebra $[L_a, L_b] = \epsilon_{abc} L_c$ $(a, b, c = 1, 2, 3)$ of the angular momentum in three dimensions.

The first step in the study of local properties is to note that the analogue of the multiplication table for finite groups is a set of *structure functions* $\phi(a, b)$ defined by

$$g(a)g(b) = g(\phi(a,b)),$$
$$a = a_k, \quad b = b_k, \quad \phi = \phi_k, \quad k = 1, ..., r, \tag{2.1}$$

where a, b, ϕ are the parameters. For example for the group $SU(2)$ with elements $(\cos a + \sigma \cdot \hat{\mathbf{a}} \sin a)$ where $\mathbf{a}$ is a 3-vector, $a = |\mathbf{a}|$ and $\hat{\mathbf{a}} = \mathbf{a}/a$, one finds that $\phi(a, b)$ is given by

$$\hat{\phi} \sin \phi = \hat{\mathbf{a}} \cos b \sin a + \hat{\mathbf{b}} \cos a \sin b - \hat{\mathbf{a}} \wedge \hat{\mathbf{b}} \sin a \sin b. \tag{2.2}$$

It is easy to see that the conditions (i), (ii), (iii) of section 1.1 for the existence of a group impose the following conditions on the structure functions:

(i) associativity $\Rightarrow \phi(a, \phi(b, c)) = \phi(\phi(a, b), c)$;
(ii) existence of the identity $\Rightarrow \phi(a, 0) = \phi(0, a) = a$;
(iii) existence of the inverse $\Rightarrow \phi(x, a) = 0$ and $\phi(a, y) = 0$ are solvable for x and y, and $x = y$.

13

It is well known that if a real function $f(a)$ is continuous and $f(a)f(b) = f(a+b)$ then $f(a)$ is real analytic (in fact $f(a) = \exp ka$, for some constant k) and a very deep theorem, due to Montgomery and Zippen (1955), generalizes this result to the statement that if a set of functions $\phi(a, b)$ are continuous and satisfy conditions (i) to (iii) then they are real analytic, and thus have the Taylor expansions

$$\phi^k(a, b) = a^k + b^k + c^k_{rs}\, a^r b^s + \begin{array}{l} \\[2pt] + c^k_{rst}\, a^r a^s b^t \\[6pt] + c^k_{rstu}\, a^r a^s a^t b^u \\[6pt] + e^k_{rstu}\, a^r a^s b^t b^u \\[6pt] + d^k_{rst}\, a^r b^s b^t \\[6pt] + d^k_{rstu}\, a^r b^s b^t b^u \\[6pt] \end{array} \quad \begin{array}{l} +\dots \\[6pt] +\dots \\[6pt] +\dots \\[6pt] +\dots \\[6pt] +\dots \end{array} \tag{2.3}$$

Note that the group condition (ii) above is just the condition that the expansion (2.3) should have no (nonlinear) pure a or pure b terms. Note that a left and right inverse exists (at least in the neighbourhood of the origin) by the inverse function theorem, and that the left and right inverses agree if the associative condition is satisfied. Thus the only condition that $\phi(a, b)$ in (2.1) must satisfy in order to be a structure function is the associativity condition.

The key to the study of Lie groups is to express the associativity condition by means of a differential equation, namely,

$$\frac{\partial \phi(a, b)}{\partial b} = u(\phi)\, v(b), \quad u(a) = \left(\frac{\partial \phi(a, b)}{\partial b}\right)_{b=0}, \quad v(b) = u(b)^{-1}, \tag{2.4}$$

where u and v are $r \times r$ matrices and u is called the *tangent* matrix. To establish (2.4) one differentiates the associativity condition (i) with respect to the right-hand variable c to obtain

$$\frac{\partial \phi(\phi(a, b), c)}{\partial c} = \frac{\partial \phi(a, \phi(b, c))}{\partial \phi(b, c)} \cdot \frac{\partial \phi(b, c)}{\partial c}, \tag{2.5}$$

and then sets $c = 0$. Note that (2.5) implies that an alternative definition for $v(a)$ is $(\partial \phi(a, b)/\partial b)_{\phi=0}$. The converse of (2.4) is also true, namely, that every $\phi(a, b)$ satisfying (2.4) satisfies (i) (see exercise 2.1).

The next step is to find the conditions under which the key equation (2.4) is integrable. For this one applies to (2.4) the usual integrability conditions for partial differential equations, namely

$$\frac{\partial}{\partial b_m}(u(\phi)\, v(b))^k_n = \frac{\partial}{\partial b_n}(u(\phi)\, v(b))^k_m. \tag{2.6}$$

Then using (2.4) again, the re-arranging terms, one sees that (2.6) is equivalent to

$$\left[\frac{\partial v_n^k(b)}{\partial b_m} - \frac{\partial v_m^k(b)}{\partial b_n}\right] u_t^m(b)\, u_s^n(b) = \left[\frac{\partial u_t^q(\phi)}{\partial \phi_p}\, u_s^p(\phi) - \frac{\partial u_s^q(\phi)}{\partial \phi_p}\, u_t^p(\phi)\right] v_q^k(\phi) = f_{ts}^k,$$

(2.7)

where the f_{ts}^k must be constants because the independent variables b and ϕ are separated. These constants are known as the structure constants, and by setting $b = 0$ in (2.7) they are easily identified as

$$f_{rs}^k = c_{rs}^k - c_{sr}^k,$$

(2.8)

where $\qquad c_{rs}^k = (\partial^2 \phi / \partial a_r\, \partial b_s)_{a-b-0}$

(i.e. as the antisymmetric part of the coefficient of the second degree term in the expansion (2.3)). The two equations in (2.7) are equivalent because of the relation $uv = 1$ and they may be summed up in the simpler form

$$[I_r, I_s] = f_{rs}^t\, I_t, \quad I_r = u_r^k(a)\frac{\partial}{\partial a^k},$$

(2.9)

where the I_r are called the infinitesimal generators.

The integrability conditions (2.7) are, of course, themselves partial differential equations and so require further integrability conditions. However, in contrast to the key equation (2.4), the integrability conditions (2.7) are self-contained, in the sense that the derivatives of the *u*s are expressed in terms of the *u*s themselves. Hence one would expect the integrability process to stop at the next step, and so it does. From the general theory of partial differential equations, the integrability condition for the curl operator is obtained by differentiating with respect to a_k and setting to zero the cyclic sum with respect to i, j, k (just as in Maxwell's equations the integrability condition for $\nabla \times \mathbf{A} = \mathbf{B}$ is $\nabla \cdot \mathbf{B} = 0$). Applying this procedure to (2.7), and using (2.4) and the invertibility of the *v*s, one obtains the (Jacobi) condition

$$f_{ls}^k f_{mn}^s + f_{ms}^k f_{nl}^s + f_{ns}^k f_{lm}^s = 0 \quad (f_{mn}^k = -f_{nm}^k).$$

(2.10)

This is the final integrability condition and is purely algebraic.

Collecting all these results, one sees that for every local Lie group the structure constants (antisymmetric part of the c_{rs}^k in (2.3)) must satisfy the Jacobi relation (2.10). Conversely, for every set of constants f_{rs}^k which are antisymmetric in r and s and satisfy the Jacobi relation, (2.7) and (2.4) can be integrated to obtain structure functions for a local Lie group. This is Lie's theorem. The only question left open is the uniqueness of the relationship between the structure constants and the local group, and this will be discussed in section 2.6.

2.2 Differential forms

Before proceeding to consider the uniqueness problem, it is convenient to consider at this point some differential consequences of the key equation (2.4). First, if (2.4) is written in the form

$$\delta\phi(a, b) = \delta b, \tag{2.11}$$

where $\delta b \equiv v(b)\,\mathrm{d}b$, it shows that the differentials δb are invariant with respect to left group multiplication with a fixed element ($g(a)$ in this case), and using this result it is easy to construct the left-invariant measure and line element promised in section 1.2, namely

$$\mathrm{d}\mu(a) = \delta a_1\,\delta a_2\ldots\delta a_r = (\det v(a))\,\mathrm{d}a_1\,\mathrm{d}a_2\ldots\mathrm{d}a_r, \tag{2.12}$$

and

$$\mathrm{d}s^2 = \eta_{ij}\,\delta a^i \delta a^j = g_{ij}(a)\,\mathrm{d}a^i\,\mathrm{d}a^j, \tag{2.13}$$

where

$$g_{ij}(a) = \eta_{rs}\,v_i^r(a)\,v_j^s(a),$$

and η_{ij} is any constant, positive, metric. The corresponding right-invariants are constructed with $\tilde{u}(a)$ where

$$\tilde{u}(a) = \left(\frac{\partial\phi(b, a)}{\partial b}\right)_{b=0}. \tag{2.14}$$

Second, if one lets $b = \delta a$ in the key equation (2.4) one obtains

$$\phi(a, \delta a) = a + u(a)\,\delta a = a + \mathrm{d}a \tag{2.15}$$

and hence

$$g(a + \mathrm{d}a) = g(a)\,g(\delta a) \quad (\text{not } g(a)\,g(\mathrm{d}a)). \tag{2.16}$$

In particular the derivative of a finite group element is

$$\frac{\partial g(a)}{\partial a^k} = g(a)\,v_k^s(a)\left(\frac{\partial g(b)}{\partial b^s}\right)_{b=0}, \tag{2.17}$$

a result that will be useful in discussing the Higgs mechanism in chapter 8.

2.3 Parameter transformations

Up to this point the coordinates a_k in the parameter space of the Lie group have not been specified, except to say that they should be locally Euclidean and that $\phi(a, b)$ should be real analytic. This was possible because the integrability conditions are covariant with respect to coordinate transformations. However, to define the group more precisely, and especially to investigate the uniqueness of the relation between the Lie group and the structure constants, one must consider the role of parameter transformations. The permitted parameter transformations are non-singular transformations of the form $a_k \to a'_k = f_k(a_j)$, where the $f_k(a_j)$ are real analytic, and it is convenient to subdivide such transformations into two kinds, namely,

(a) linear transformations $a_k \to M_{ks} a_s$, M_{ks} non-singular, and
(b) nonlinear transformations whose Jacobian is unity at $a = 0$, $a_k \to a_k + O(a^2)$.

It is evident that any permitted transformation is a combination of these two. One sees at once that $\phi^k(a, b)$, $u_r^k(a)$, c_{rs}^k, f_{rs}^k and all other indexed quantities behave as tensors of obvious rank with respect to the linear transformations (a). On the other hand, with respect to the nonlinear transformations (b), the tangent matrices transform only as vectors, and the structure constants are actually *invariant*,

$$u_r^k(a) = \frac{\partial a^k}{\partial b^j} u_r^j(b), \quad f_{jk}^i(a) = f_{jk}^i(b). \qquad (2.18)$$

This is because the $u(a)$ and f are defined by differentiation at the origin.

There exists for each group a natural set of parameters called normal parameters, which completely eliminates the freedom (b). They are defined by the condition

$$u_k^s(b) b^k = b^s, \qquad (2.19)$$

and it is easy to see that they exist and are unique up to linear transformations (a) by noting that (2.19) is equivalent to the statement that the completely symmetric part of the upper line of coefficients c_{rs}^k, c_{rst}^k, c_{rstu}^k, ... in (2.3) vanishes, and the completely symmetric coefficients in the expansion of the transformation $a \to b = a + O(a^2)$ can always be chosen so that this is true. Furthermore, the choice is unique. The group-theoretical meaning of the normal parameters is that each one describes a one-parameter subgroup of the original group, with the additive multiplication law $\phi(b, b') = b + b'$. This can be seen by noting that $\phi = b + b'$ is the

unique solution of the key equation (2.4) whenever (2.19) is true and all
the parameters except one are set equal to zero. The geometrical meaning of
the normal coordinates is that they are geodesic coordinates with respect
to the left-invariant metric (2.13) (Pontryagin, 1966). These interpretations
imply that the one-parameter subgroups and geodesics coincide (at least
locally) and an elegant direct proof of this result can be found in Sternberg
(1983).

The normal parameters can be used to show that the structure constants
determine the structure functions uniquely, up to coordinate transformations, because (2.19) can be added to the curl equation (2.7) to obtain

$$\frac{d}{ds}(sv_t^k(b)) = \delta_q^k + f_{pq}^k(sv_t^p(b))\left(\frac{b^q}{s}\right), \tag{2.20}$$

where $s^2 = b^q b^q$, and since this is an ordinary differential equation with a
given initial condition, it uniquely determines the $v(b)$. The $\phi(a, b)$ are then
determined uniquely by the key equation (2.4).

2.4   Local Lie groups and Lie algebras

Since the structure constants determine the Lie groups up to parameter
transformations, and are themselves invariant with respect to transformations of the kind (b), there is a one–one correspondence between Lie
groups and the equivalence classes of structure functions

$$f \sim f' \Leftrightarrow f_{st}^r = (M^{-1})_i^r M_s^j M_t^k f_{jk}^{\prime i}, \tag{2.21}$$

where M are the linear transformations (a). A natural way to characterize
such equivalence classes is by *Lie algebras*, where Lie algebras are defined
as linear vector spaces L with multiplication law $[L, L] \to L$ such that

$$[X, Y] + [Y, X] = 0, \quad [X[Y, Z]] + \text{cyclic} = 0, \quad X, Y, Z \in L. \tag{2.22}$$

To see this one notes that if the elements of L are expanded in a basis
σ_k, $X = x^k \sigma_k$, etc., then (2.22) takes the form

$$[\sigma_s, \sigma_t] = f_{st}^k \sigma_k, \tag{2.23}$$

where the f_{st}^k satisfy exactly the Jacobi relations (2.10) and the transformations (2.21) correspond exactly to changes of basis.

The one–one correspondence between Lie algebras and the equivalence
classes of structure constants shows that there is a one–one correspondence
between Lie algebras and local Lie groups, and this correspondence can be

expressed in an explicit and practical way by exponentiating the Lie algebra, i.e. by writing

$$\gamma(a) = \exp a \cdot \sigma, \quad a \cdot \sigma \equiv a^k \sigma_k, \tag{2.24}$$

where a^r are the parameters and σ_r the base elements. Then the Baker–Campbell–Hausdorff (BCH) formula (Varadarajan, 1974) for the product of exponentials of associative operators

$$\exp A \exp B = \exp\{A + B + \tfrac{1}{2}[A, B] + \tfrac{1}{12}[A[AB]] + \tfrac{1}{12}[B[BA]] + \ldots\}, \tag{2.25}$$

shows that

$$\gamma(a)\gamma(b) = \gamma(\phi(a, b)),$$

where

$$\phi^p(a, b) = a^p + b^p + f^p_{ij} a^i b^j + \tfrac{1}{6} f^p_{iq} f^q_{jk}(a^i a^j b^k + b^i b^j a^k) + \ldots. \tag{2.26}$$

From this equation one sees that the $\gamma(a)$ are elements of a local Lie group, and since the structure constants are the same as the original group, the two groups must be locally isomorphic. Thus the $\gamma(a)$ can be used as an explicit form of the $g(a)$ and have the advantage that all properties deduced up to now (associativity, existence of the inverse, existence of one-parameter subgroups) are displayed explicitly. Furthermore, the parameters are normal, and the manner in which the $\phi(a, b)$ are completely determined by the structure constants can be seen by inspection of (2.26). Note in particular that $g^{-1}(a) = g(-a)$. Finally, by writing (2.15) in the exponential form

$$\exp[a + u(a)\delta] \cdot \sigma = \exp a \cdot \sigma \exp \delta \cdot \sigma, \tag{2.27}$$

and using the BCH formula one obtains a closed form for the $u(a)$, namely

$$u(a) = \sum_{n=0}^{\infty} c_n (a \cdot f)^n = (a \cdot f)(1 - e^{-a \cdot f})^{-1}, \tag{2.28}$$

where f is the matrix $f^t_s = a^r f^t_{rs}$, c_n are the coefficients of the terms in the BCH formula which are linear in B, and the function on the right-hand side is derived in exercise 2.3. Equation (2.28) implies, of course, that

$$v(a) = (1 - e^{-a \cdot f})/(a \cdot f), \quad \tilde{u}(a) = u(-a) = u(a) e^{-a \cdot f}. \tag{2.29}$$

In practice the exponentiation of the algebra is by no means confined to a small neighbourhood of the origin. Indeed, for compact groups the coincidence of one-parameter subgroups and geodesics, and the general result that compact manifolds are geodesically complete, means that every element in the group can be written as an exponential. This is not true for non-compact groups (for $SL(2, R)$ for example it is easy to see that the span of $\exp(a \cdot \sigma)$ contains only those elements with trace ≥ 2 and the single element -1) and, of course, even for compact groups it may not be the most convenient way to write the elements (for example, for $SU(2)$ the well-known form $\exp \psi \sigma_3 \exp \theta \sigma_2 \exp \phi \sigma_3$ is often preferred). But the result indicates the scope of the exponential formalism, and one could even go so far as to *define* a local Lie group as an exponential Lie algebra. Then the Montgomery–Zippen (MZ) and Lie theorems would be the statements that every locally Euclidean group is real analytic, and every local real analytic group is a local Lie group, respectively.

2.5 Conjugation and adjoint representation

For topological groups in general the operation of conjugation preserves the neighbourhood of the identity ($g(b) \to e$ implies that $g(a)\,g(b)\,g^{-1}(a) \to e$ by continuity) and for Lie groups this means that, with respect to conjugation, the infinitesimal parameters, Δb say, transform *linearly*,

$$g(a)\,g(\Delta b)\,g^{-1}(a) = g(A(a)\,\Delta b), \qquad (2.30)$$

where $A(a)$ is an $r \times r$ matrix. Furthermore, by letting $g(a) \to g(c) = g(h)\,g(a)$ in (2.30) one sees that $A(a) \to A(c) = A(h)\,A(c)$ and thus the $A(a)$s constitute a linear representation of the group (see chapter 4). This representation is called the *adjoint* representation, and although it is not quite faithful (one-to-one), it is faithful modulo the centre Z,

$$A(a) = 1 \Rightarrow [g(a), g(\Delta b)] = 0 \Rightarrow g(a) \in Z, \qquad (2.31)$$

(assuming that the group is connected). In more formal language, Z is the kernel of the homomorphism $G \to A$ and A is a faithful representation of the quotient group G/Z. On account of this, the quotient group G/Z with no centre is often called the *adjoint group*.

An explicit form for A can be found by using (2.16) in the forms

$$g(a)\,g(\Delta b) = g(a + u(a)\,\Delta b), \quad g(\Delta b)\,g(a) = g(a + \tilde{u}(a)\,\Delta b), \quad (2.32)$$

for left and right multiplication respectively. From these equations and (2.30) one easily sees that

$$A(a) = \tilde{v}(a)\,u(a). \qquad (2.33)$$

All these results may be seen explicitly and $A(a)$ computed in explicit form by using the exponentiated algebra (2.24). One sees that

$$\gamma(a)\,\gamma(\Delta b)\,\gamma^{-1}(a) = e^{a \cdot \sigma}\,\Delta b \cdot \sigma\,e^{-a \cdot \sigma}$$
$$= \sum_{n=0}^{\infty} \frac{1}{n!}\,[a \cdot \sigma[a \cdot \sigma[\ldots[a \cdot \sigma, \Delta b \cdot \sigma]\ldots]]], \quad (2.34)$$

and hence that $A(a)$ may be written as

$$A(a) = \sum_{n-0}^{\infty} \frac{1}{n!}(a \cdot f) = \exp(a \cdot f). \tag{2.35}$$

where $(a \cdot f)_s^t$ is the matrix $a^q f_{qs}^t$ defined earlier. Note that (2.33) and (2.35) agree with the expressions (2.28) and (2.29) for $u(a)$ and $v(a)$.

From (2.33) it follows that the left- and right-invariant metrics and line elements (2.12) and (2.13) are related by

$$\mathrm{d}\mu_L(a) = \det A(a)\, \mathrm{d}\mu_R(a), \quad g_L(a) = A(a)\, g_R(a)\, A^t(a). \tag{2.36}$$

Since $\det A(a)$ is a one-dimensional representation of the group one sees that the left and right measures are the same for all groups whose one-dimensional representations are trivial (notably the semi-simple groups, compact or non-compact) but that the left and right line elements are the same only for those groups whose adjoint representations are orthogonal. It will be seen (section 3.4) that such groups must be abelian or compact (or a mixture of the two).

Exercises

2.1. Show that the functions $f(x) = \phi(x, \phi(b, c))$ and $h(x) = \phi(\phi(x, b), c)$ both satisfy the differential equation $\partial y / \partial x = u(y)v(\phi(y))$ if ϕ satisfies (2.4), and deduce that any ϕ which satisfies (2.4) satisfies the associativity condition (i).

2.2. Let $\{\gamma_r, \gamma_s\} = \delta_{rs}, r, s = 1, \dots, 4$ be the (hermitian) Dirac matrices in Euclidean space, and $\gamma_5 = \gamma_1 \gamma_2 \gamma_3 \gamma_4$. Show that the three Lie algebras $\{[\gamma_r, \gamma_s]\}$, $\{[\gamma_r, \gamma_s]$ plus $[\gamma_r, \gamma_5]\}$ and $\{[\gamma_r, \gamma_s], [\gamma_r, \gamma_5], \text{plus } i\gamma_r, i\gamma_5\}$ exponentiate to form the groups $SU(2) \times SU(2)$, $Sp(4)$, and $SU(4)$, respectively.

2.3. When the terms of order B^2 and higher are neglected in the BCH formula (2.25) it evidently reduces to

$$\exp A \exp B = \exp\{A + B + \tfrac{1}{2}[A, B] + \tfrac{1}{12}[A[AB]] + \dots + c_n[A[A \dots [A, B] \dots]]\}.$$

By using the matrices $\tfrac{\alpha}{2}\!\left(\begin{smallmatrix}1 & 0\\0 & -1\end{smallmatrix}\right)$, $\left(\begin{smallmatrix}0 & 1\\0 & 0\end{smallmatrix}\right)$ for A and B show that the coefficients c_n satisfy the relation

$$\sum_{n-0}^{\infty} c_n \alpha^n = \left(\frac{\alpha}{1 - e^{-\alpha}}\right)$$

used in section 2.4.

3
Lie algebras

3.1 General properties of Lie algebras

In chapter 2 it was shown that there is a one-to-one correspondence between local Lie groups and Lie algebras. This correspondence is very useful because, for many purposes, such as classification and representation theory, it is much easier to deal with the Lie algebras. Hence it is worthwhile to consider Lie algebras in their own right and that is the purpose of the present chapter.

As already noted in chapter 1 Lie algebras may be defined as (finite-dimensional) vector spaces equipped with a multiplication law $L \times L \to L$ denoted by $[X, Y]$ which satisfies the conditions

$$[X, Y] = -[Y, Z], \quad [X[Y, Z]] + [Y[Z, X]] + [Z[X, Y]] = 0, \quad X, Y, Z \in L. \tag{3.1}$$

Note that the multiplication law is neither commutative nor associative. The product $[X, Y]$ will be referred to as the *commutator* of X and Y because in many cases, in particular Lie algebras of matrices or of linear differential operators, the product $[X, Y]$ is a commutator of the form $XY - YX$ where XY is another, associative, multiplication. In that case (3.1) is automatically satisfied. However, $[X, Y]$ need not necessarily be of the form $XY - YX$, and a well-known counter-example is the case when $[X, Y]$ is the Poisson bracket of X and Y. On the other hand, it can be shown that every Lie algebra has a faithful matrix representation and so there is no great loss in generality in visualizing $[X, Y]$ as a commutator of matrices.

The isomorphism between the local Lie group and Lie algebra expressed by the exponential form (2.24) allows the properties of the local groups to be converted into properties of the algebras (and conversely), and corresponding to the definitions for groups introduced in chapter 1 one obtains the following definitions for Lie algebras:

(a) A subalgebra K of a Lie algebra L is a subspace of L which closes

22

with respect to commutation, i.e. $X, Y \in K \Rightarrow [X, Y] \in K$. (Symbolically, $[K, K] \subset K$.)

(b)   An invariant subalgebra K of a Lie algebra is a subalgebra which is invariant with respect to commutation by all elements of L, i.e. $Y \in K, X \in L \Rightarrow [X, Y] \in K$. (Symbolically, $[L, K] \subset K$.)

(c)   A Lie algebra $L = M + K$ is the direct sum of two Lie algebras M and K if it is the vector sum and all the elements of M commute with all the elements of K, i.e. $X \in M, Y \in K \Rightarrow [X, Y] = 0$. (Symbolically, $L = M + K$ where $[M, K] = 0$.)

(d)   A Lie algebra $L = M \wedge K$ is a semi-direct sum of two subalgebras M and K if one of the subalgebras (K say) is an invariant subalgebra. (Symbolically, $L = M \wedge K$ if $[K, K] = K$, $[M, M] = M$ and $[M, K] = K$.) Alternatively, one may say that an invariant subalgebra K and its complement $M = L - K$ form a semi-direct sum if, and only if, M is a subalgebra. Note also that a semi-direct sum becomes a direct sum if, and only if, both subalgebras are invariant.

(e)   An abelian Lie algebra is one for which all the elements commute, i.e. $[X, Y] = 0$ all $X, Y \in L$. In any Lie algebra L the set of elements that commute with all the elements of L forms an abelian subalgebra, called the centre Σ, and $L = L_0 + \Sigma$, where L_0 has no centre.

(f)   A Lie algebra which contains no invariant abelian subalgebra is called semi-simple, and a semi-simple algebra which contains no invariant subalgebra is called simple. It will be seen in the next section that semi-simple algebras are direct sums of simple ones.

The Lie algebras corresponding to the twelve Lie groups listed in section 1.3 are easily seen to be the following:

(1) $gl(n, c/r) = $ all complex/real $n \times n$ matrices.

(2) $sl(n, c/r) = $ all complex/real traceless $n \times n$ matrices.

(3) $d(n, c/r/u) = $ all   complex/real/pure-imaginary   diagonal $n \times n$ matrices.

(4) $t(n, c/r) = $ all complex/real upper-triangular $n \times n$ matrices.

(5) $t_0(n, c/r) = $ all complex/real strictly upper-triangular $n \times n$ matrices (no diagonal terms).

(6) $u(n) = $ all complex skew-hermitian $n \times n$ matrices.

(7) $su(n) = $ all complex skew-hermitian traceless $n \times n$ matrices.

(8) (9) $o(n, c/r) = so(n, c/r)$ all complex/real skew-symmetric $n \times n$ matrices.

(10) $sp(2n, c/r/u) = $ all complex/real/skew-hermitian, symplectic $n \times n$ matrices ($L^t J + J L = 0$).

(11) $su(p, \cdot q) =$ all pseudo-skew-hermitian $(L^\dagger G + GL = 0)$ traceless $n \times n$ matrices.

(12) $so(p, q) =$ all pseudo-skew-symmetric $(L^\dagger G + GL = 0)$ $n \times n$ matrices.

In all these cases, commutation is understood to mean ordinary matrix commutation.

3.2 The Cartan metric

A very useful concept in the study of Lie algebras is the symmetric Cartan metric g_{ab}, defined as
$$g_{ab} = f^d_{ac} f^c_{bd}, \qquad (3.2)$$

where $X = x \cdot \sigma = x^a \sigma_a$, $a = 1, \dots, r$, for any basis σ_a. This metric can be used as a lowering operator for the indices on the structure constants.

$$f_{abc} = g_{ad} f^d_{bc}, \qquad (3.3)$$

and it is easily verified from the Jacobi relation that the lower-index structure constants f_{abc} are antisymmetric in *all* indices, not just b and c. The metric g_{ab} is, of course, zero for abelian algebras and in general it may be degenerate. In non-degenerate cases one can define the inverse matrix g^{ab} as
$$g^{ab} g_{bc} = \delta^a_c \quad (\det g \neq 0), \qquad (3.4)$$

and thus recover the f^a_{bc} from the f_{abc}.

Note that in general the restriction of the Cartan metric to a subalgebra is not the same as the metric for the subalgebra,
$$g_{ij} = f^b_{ia} f^a_{jb} \neq f^k_{il} f^l_{jk} \quad \text{(in general)}, \qquad (3.5)$$

where i, j, k are the base indices of the subalgebra, but that it is the same if the subalgebra is invariant (since then the indices a and b in (3.5) are restricted to the subalgebra). An immediate consequence of this result is that the metric is zero on any abelian invariant subalgebra and hence that a non-degenerate metric $(\det g \neq 0)$ implies semi-simplicity. The converse is also true (i.e. the metric is non-degenerate if, and only if, the algebra is semi-simple) but will not be proved here. Proofs are given in most of the books in the bibliography (e.g. Gilmore, 1974).

The metric g_{ab} may be used to define a (possibily degenerate) inner product for the Lie algebra, namely,
$$(X, Y) \equiv (x, y) = g_{ab} x^a y^b \qquad (3.6)$$

where $X = x^a I_a$, $Y = y^a I_a$, and because of the complete antisymmetry of the f_{abc} the inner product has the useful cyclic property
$$(X, [Y, Z]) = (Y, [Z, X]) = (Z, [X, Y]) \ (= f_{abc} x^a y^b z^c). \qquad (3.7)$$

One use of this property is to show that semi-simple algebras are direct sums of simple algebras, as follows: Let I be an invariant subalgebra of a semi-simple Lie algebra and C its complement with respect to the Cartan metric, i.e. $(C, I) = 0$. Then from (3.7),

$$([C,I],I) = (C,[I,I]) = (C,I) = 0, \tag{3.8}$$

because I is a subalgebra, and

$$([C,I],C) = (I,C) = 0, \tag{3.9}$$

because I is invariant. Thus $[C,I]$ is orthogonal to all elements and since the metric is not degenerate $[C,I]$ must be zero. This means that every invariant subalgebra commutes with its complement, which, in turn, means that the algebra can be decomposed into a direct sum of algebras with no invariant subalgebra, i.e. into a direct sum of simple algebras,

$$G = G_1 + G_2 + \ldots + G_k, \tag{3.10}$$

as required.

3.3   The adjoint representation of the algebra

The adjoint representation of a Lie group has already been discussed in chapter 1, from which it can be written as

$$A(x) = \exp T(x), \tag{3.11}$$

where $T(x)^r_s \equiv x^t f^r_{ts}$, and it acts on the space of the Lie algebra by conjugation. Accordingly, the generators $T(x)$ of A should form a representation of the algebra on itself by conjugation, i.e.

$$T(x)y = y', \tag{3.12}$$

where $[X, Y] = Y'$ and $X = x \cdot \sigma$, $Y = y \cdot \sigma$, and this is easily verified using the Jacobi identity. The representation $T(x)$ is called the adjoint representation of the algebra and is a finite-dimensional representation of the algebra on itself in much the same way that the regular representation of a finite group is a representation of the group on itself.

Since the adjoint representation $A(x)$ of the group is faithful modulo the centre Z the adjoint representation $T(x)$ of the algebra is faithful modulo the centre Σ of the algebra. Thus if the group centre Z is discrete, the centre of the algebra Σ vanishes, and $T(x)$ is a faithful representation of the algebra. In particular, the adjoint representation $T(x)$ is faithful for semi-simple algebras, since if $T(x)$ were zero for any x, then $x \cdot \sigma$ would commute with the whole algebra and thus form an abelian invariant (central) subalgebra. Note that in the adjoint representation the Cartan inner product may be realized as a trace

$$(X, Y) = (x, y) = \text{tr}\,(T(x)\,T(y)) \tag{3.13}$$

and that on account of the cyclic property (3.7) of the inner product,

$$(x, T(y)z) + (T(y)x, z) = 0, \tag{3.14}$$

which shows that the adjoint representation of the algebra is skew-symmetric, and the adjoint representation of the group is orthogonal, with respect to this inner product. It will be recalled, however, that the Cartan inner product may be indefinite, or even degenerate.

3.4 Compact Lie algebras

For gauge theories, the case of compact Lie groups is of special interest, since, with the exception of gravitation, the gauge theories in use at present are based on compact Lie groups. For this reason it is desirable to express the compactness of a group in terms of its Lie algebra, and to do this one uses the following result: The adjoint representation of a Lie group is orthogonal if, and only if, the adjoint group G/Z is compact. This result is the one anticipated in section 2.5 to state that the Riemannian line element (2.13) is left- and right-invariant only if G/Z is compact, and it may be shown as follows:

Suppose first that A is orthogonal. Then, since it is real, it is unitary. But then since A is a faithful representation of G/Z the group G/Z is unitary and hence compact. Conversely, if G/Z is compact, the positive-definite inner product

$$\langle f, g \rangle = \int d\mu(a)\,(f, A^t(a)\,A(a)g), \quad a \in G/Z, \tag{3.15}$$

exists and it is easy to see that $A(b)$ is orthogonal with respect to it.

Applying this result to the adjoint representation of the Lie algebra one sees that it is (real) skew-symmetric if and only if G/Z is compact. In that case the structure constants f^t_{sq} are antisymmetric in t and q as well as s and q

$$f^t_{sq} = -f^q_{st} \quad (\text{and}\, f^t_{sq} = -f^t_{qs}), \tag{3.16}$$

and hence are totally antisymmetric in t, s, q (without lowering). The condition (3.16) is the required intrinsic condition for compactness.

One immediate consequence of (3.16) is that for a compact Lie algebra the Cartan metric is negative,

$$g_{sq} = f^t_{sp}f^p_{qt} = -f^t_{sp}f^t_{qp} \leqslant 0, \tag{3.17a}$$

and is zero only on a central part

$$g_{sq}x^q = 0 \Rightarrow f^t_{sq}x^q = 0 \Rightarrow [X \cdot I, y \cdot I] = 0 \quad \text{for all } y. \qquad (3.17b)$$

Since these statements are basis-independent and an algebra for which g_{ab} is not degenerate is semi-simple, one sees at once that every compact Lie algebra is a direct sum of a semi-simple algebra L_S and an abelian algebra L_A, and that the Cartan metric is negative-definite and zero on the respective parts,

$$g(L_S) < 0, \quad g(L_A) = 0, \qquad (3.18)$$

where $L_C = L_S + L_A$. Finally, since a semi-simple Lie algebra is a direct sum of simple Lie algebras, and abelian Lie algebras are evidently a direct sum of one-dimensional Lie algebras, one obtains the following structure theorem for compact Lie algebras:

THEOREM

Any compact Lie algebra is the direct sum of irreducible compact Lie algebras, where irreducible means either simple or one-dimensional.

Correspondingly, any compact Lie group is the direct product of simple compact Lie groups and one-parameter compact Lie groups $U(1)$, modulo a discrete centre. Note that, on account of Schurs lemma, the inner product (3.15) and the Cartan inner product must coincide (up to a constant) on the irreducible parts.

This structure theorem for compact Lie groups is of great importance for physics because it shows that the only compact gauge fields are combinations of the Maxwell fields associated with $U(1)$ and the Yang–Mills fields associated with simple compact Lie algebras. It will be seen later that there is just one fundamental coupling constant for each irreducible algebra in the direct sum reduction of the theorem.

3.5   Cartan canonical form of simple compact Lie algebras

For any Lie algebra it is desirable to use the freedom of linear transformations in order to reduce the number of non-zero structure constants f^a_{bc}, and in the case of compact simple algebras a reduction of this kind is used to cast the algebra into a simple form which is known as the Cartan canonical form. The Cartan form is the generalization of the well-known form

$$[I_3, I_\pm] = \pm I_\pm, \quad [I_+, I_-] = I_3, \qquad (3.19)$$

of the angular-momentum algebra, and it is particularly useful for physics because the commuting elements (generalization of I_3) are usually good quantum numbers. For example, they define the various charges (electric, strange, charmed, coloured) of the elementary particles.

The technique for constructing the Cartan form is to use the adjoint representation $T(x)$ instead of the algebra itself and to complexify it, where complexification means letting the coefficients x in $x \cdot I$ and the Cartan inner product (x, y) become complex. Note that there is no loss of generality in using the adjoint representation since for compact simple algebras it is faithful.

The first step in the reduction is to find a maximal set of linearly independent commuting elements $T(H_i)$, $i = 1, ..., l$, in the algebra. The number l of such elements is called the *rank* of the algebra and it can be shown that it is independent of the choice of the $T(H_i)$. For convenience it is usual to orthonormalize the $T(H_i)$ with respect to the Cartan metric and thus we have

$$[T(H_i), T(H_j)] = 0, \quad (T(H_i), T(H_j)) \equiv \mathrm{tr}\{T(H_i) T(H_j)\} = \delta_{ij}. \quad (3.20)$$

Note that since the inner product in (3.20) is positive the H_i are not elements of the original real Lie algebra but elements of that algebra multiplied by $i = \sqrt{-1}$. The i is inserted in order to make the $T(H_i)$ hermitian, in which case the H_i can be identified as the charge quantum numbers mentioned above. The abelian subalgebra spanned by the H_i is called the Cartan subalgebra. For the Lie algebras of the classical groups the Cartan algebra is simply the algebra of all diagonal matrices compatible with the group conditions (unimodular, orthogonal, symplectic).

The next step is to diagonalize the commuting matrices $T(H_i)$. Since, before diagonalization, these matrices are just $T(H_i) = i f_{ib}^a$, which are pure imaginary antisymmetric, their non-zero eigenvalues must come in pairs of equal magnitude and opposite sign. Thus one can write

$$T(H_i) V_\alpha = \alpha^{(i)} V_\alpha, \quad V_\alpha^* = V_{-\alpha}, \quad \alpha \neq 0, \quad (3.21)$$

where V_α are the corresponding eigenspaces. Furthermore, the V_α are orthogonal for $\alpha + \beta \neq 0$, so

$$(V_\alpha, V_\beta) = 0 \text{ unless } \alpha + \beta = 0. \quad (3.22)$$

But now, since the action of the adjoint representation is by commutation, and the metric with respect to which the $T(H_i)$ were originally antisymmetric is the Cartan one, the equations (3.21) can be converted into the commutation relations

$$[T(H_i), T(V_{\pm\alpha})] = \pm \alpha T(V_{\pm\alpha}),$$

where
$$\operatorname{tr} T(V_\alpha)\, T(V_\beta) = 0 \text{ for } \alpha+\beta \neq 0, \quad T^\dagger(V_\alpha) = T(V_{-\alpha}). \quad (3.23)$$

The V_α evidently span the $(r-l)$-dimensional complement of the Cartan algebra in the Lie algebra (which means, incidentally, that $r-l$ must be even). Equation (3.23) already resembles the prototype equation (3.19), but to complete the reduction, some more steps are necessary.

The most important step is to show that the spaces V_α are actually one dimensional. To show this let $E_{-\alpha}$ be a fixed element of $V_{-\alpha}$ and let E_α denote any element of V_α such that $(E_{-\alpha}, E_\alpha) = c \neq 0$. Such an element must exist since otherwise, from (3.23), $E_{-\alpha}$ would be orthogonal to everything and hence, by the positivity of the metric, it would be zero. From the Jacobi relation one sees that

$$[T(E_\alpha), T(E_{-\alpha})] = x^{(i)} T(H_i), \quad (3.24)$$

where $x^{(i)} = c\alpha^{(i)}$. Now consider the space spanned by $\{E_{-\alpha}, \alpha \cdot H, E_\alpha, E_{2\alpha}, ..., E_{k\alpha}\}$ where k is the largest multiple of α that can occur. From the Jacobi relation it follows that this subspace is invariant with respect to commutation with $T(E_{-\alpha})$, $T(E_\alpha)$ and $T(\alpha \cdot H)$. In that case the commutation relation can be restricted to this subspace. Restricting it, and taking the trace on the subspace one obtains

$$0 = c|\alpha|^2(-1 + \dim V_\alpha + 2 \dim V_{2\alpha} + ... + k \dim V_{k\alpha}), \quad (3.25)$$

which shows that $\dim V_\alpha = 1$ as required (and incidentally that $k = 1$).

The final step is to note that, because of the Jacobi relation, one has

$$[T(H_i)[T(E_\alpha), T(E_\beta)]] = (\alpha+\beta)^{(i)} [T(E_\alpha), T(E_\beta)] \quad (3.26)$$

and since the eigenspaces are one-dimensional, the commutator on the right-hand side of (3.26) must be a multiple of $T(E_{\alpha+\beta})$. With suitable normalization ($c = 1$) one therefore has

$$[T(E_\alpha), T(E_\beta)] = N_{\alpha\beta}\, T(E_{\alpha+\beta}), \quad \operatorname{tr}(T(E_\alpha)\, T(E_\beta)) = \delta_{\alpha+\beta,\, 0}, \quad (3.27)$$

where $N_{\alpha\beta}$ is a coefficient depending only on α and β (and may, of course, be zero).

Combining all these results, and converting them from the adjoint representation to the Lie algebra itself, one obtains finally the relations

$$\left.\begin{array}{l} [H_i, H_j] = 0, \quad [H_i, E_{\pm\alpha}] = \pm\alpha^{(i)} E_{\pm\alpha}, \quad (i,j = 1, ..., l) \\ [E_\alpha, E_{-\alpha}] = \alpha^{(i)} H_i, \quad [E_\alpha, E_\beta] = N_{\alpha\beta} E_{\alpha+\beta}, \quad (\alpha+\beta \neq 0), \end{array}\right\} \quad (3.28)$$

where the H_i and E_α are orthonormal, i.e.

$$\operatorname{tr}(H_i H_j) = \delta_{ij}, \quad \operatorname{tr}(E_\alpha E_\beta) = \delta_{\alpha+\beta,\, 0}, \quad E_\alpha^\dagger = E_{-\alpha}.$$

This is the Cartan canonical form of the algebra and is the required generalization of (3.19). The reduction in the number of structure constants achieved by (3.28) is quite substantial, namely from $\frac{1}{2}r^2(r-1)$ three-index coupling constants f^a_{bc} to $\frac{1}{2}(r-l)\,(r-1)$ two-index constants $\alpha^{(i)}$ and $N_{\alpha\beta}$. Actually, as will be seen later, the $N_{\alpha\beta}$ are determined by the $\alpha^{(i)}$, and the $\alpha^{(i)}$ are determined in terms of l fundamental eigenvalues, so the reduction is even greater than appears at first sight.

3.6 Root diagrams and the Weyl group

The constants α in (3.28) may be thought of as vectors in an l-dimensional Euclidean space $E(l)$, with inner product $\langle\,,\,\rangle$ say. This space $E(l)$ is called the root space, and, as will now be seen, the angles between the roots and their lengths in $E(l)$ are not arbitrary. In fact they are quantized, so that the roots for each compact simple algebra form a kind of crystal (cf. fig. 1) called a root diagram.

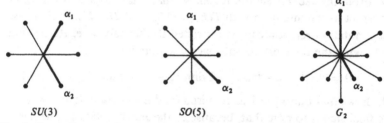

Fig. 1. Root diagrams for the simple compact Lie groups of rank two, namely, $SU(3)$, $SO(5)$ ($\approx Sp(4)$)and G_2. The primitive roots are drawn thicker and are labelled α_1 and α_2.

To obtain the quantization of root lengths and angles, one first notes that the roots form 'strings' of the form $\alpha, \alpha-\beta, \alpha-2\beta, ..., \alpha-m\beta$, where the sequence is finite and is obtained by successive commutation with $E_{-\beta}$, i.e. $E_\alpha \to [E_{-\beta}, E_\alpha] \to [E_{-\beta}[E_{-\beta}, E_\alpha]]$ and so on. By definition, $[E_\beta, E_\alpha]$ and $[E_{-\beta}, E_{\alpha-m\beta}]$ are zero. Consider now the norms N_r defined as

$$N_r^2 = \mathrm{tr}\,(T(E_{-\alpha})\,T^r(E_\beta)\,T^r(E_{-\beta})\,T(E_\alpha)), \quad r = 1, 2, ... \qquad (3.29)$$

in the space of the adjoint representation. By using the Cartan basis one easily establishes the recurrence relation

$$N_r^2 = [r\langle\alpha,\beta\rangle - \tfrac{1}{2}r(r-1)\,\langle\beta,\beta\rangle]\,N_{r-1}^2, \qquad (3.30)$$

which shows that N_r vanishes if, and only if,

$$\frac{2\langle \alpha, \beta \rangle}{\langle \beta, \beta \rangle} = (r-1) = \text{integer.} \tag{3.31}$$

But since the string ends at $r = m+1$ this condition is satisfied for $r-1 = m$.

Since the roles of α and β are interchangeable (3.31) implies that

$$\cos^2 \theta_{\alpha\beta} = \frac{\langle \alpha, \beta \rangle^2}{\langle \alpha, \alpha \rangle \langle \beta, \beta \rangle} = \tfrac{1}{4} m_1 m_2 \tag{3.32}$$

where $\theta_{\alpha\beta}$ is the angle between α and β and m_1 and m_2 are integers. Equation (3.32) is the required quantization condition. It implies that the only non-zero (acute) angles between roots are π/n where $n = 6, 4, 3, 2$ and that for $n = 3, 4, 6$ the relative lengths of the roots are fixed, in the ratios 1, . $\sqrt{2}$ and $\sqrt{3}$ respectively. Examples of all four cases occur for the simple compact groups of rank $l = 2$ and the corresponding root diagrams are shown in fig. 1.

The root diagrams of fig. 1 exhibit a high degree of symmetry. This is a general feature of root diagrams and is due to the fact that they are invariant under a (finite) group, called the Weyl group, which is generated by reflections in the planes orthogonal to the roots. To see that the weight diagrams are invariant with respect to Weyl reflexions, one simply notes that for the 'string' $\alpha, \alpha-\beta, \alpha-2\beta, ..., \alpha-m\beta$ discussed above, the end-root

$$\alpha - m\beta = \alpha - \frac{2\langle \alpha, \beta \rangle}{\langle \beta, \beta \rangle} \beta, \tag{3.33}$$

is just the reflexion of α in the plane orthogonal to the root β. Since this hold for all pairs of roots α, β the weight diagram must be invariant with respect to all such reflexions.

The Weyl group W permutes the base elements H_i of the Cartan algebra, and may even be defined by this property as follows: Let N ($=$ normalizer) and C ($=$ centralizer) be the maximal subgroups of the covering group $\tilde{G}$ which leave the Cartan algebra and each separate element of the Cartan algebra invariant. Then $C = \exp(i\alpha_s H_s)$ and W may be defined as the quotient N/C. Note that since W is only a quotient, it is not necessarily a subgroup of $\tilde{G}$ itself. For example, it is a subgroup for $SU(2n+1)$ but not for $SU(2n)$.

Finally, it might be mentioned that by using arguments similar to that leading from (3.29) to (3.31), it can be shown that the $N_{\alpha\beta}$ satisfy the relations

$$N_{\alpha\beta}+N_{\beta\gamma}+N_{\gamma\alpha}=0 \quad \text{for } \alpha+\beta+\gamma=0, \tag{3.34}$$

and

$$N_{\lambda\alpha}N_{\beta\gamma}+N_{\lambda\beta}N_{\gamma\alpha}+N_{\lambda\gamma}N_{\alpha\beta}=0 \quad \text{for } \alpha+\beta+\gamma+\lambda=0. \tag{3.35}$$

From the first relation it follows that

$$|N_{\alpha\beta}|^2 = \tfrac{1}{2}st(\beta,\beta) \tag{3.36}$$

where $\alpha-(t-1)\beta$, ..., $\alpha-\beta$, α, $\alpha+\beta$, ..., $\alpha+s\beta$, is a string, and from the second relation it follows (though not quite so easily) that the $N_{\alpha\beta}$ can be chosen to be real (even integer) and to have definite signs (see references, for example, Wybourne, 1974). From these results it follows that, up to a phase convention, the structure constants $N_{\alpha\beta}$, and hence the Lie algebras, are determined by the roots α.

3.7 Primitive roots, Dynkin diagrams and classification

As just stated, all the structure constants are determined by the roots a. On the other hand, there are $\tfrac{1}{2}(r-l)$ roots, and since they lie in an l-dimensional vector space one might expect them to be determined by a fundamental set of l roots. This is so, and the fundamental roots, called *primitive roots*, are defined as follows: Choose any ordering H_1, H_2, ..., H_l for the basis of the Cartan subalgebra and define a root $\alpha = (\alpha^{(1)}, \alpha^{(2)}, ..., \alpha^{(l)})$ to be positive if $\alpha^{(1)} > 0$ or $\alpha^{(1)} = 0$, $\alpha^{(2)} > 0$ or $\alpha^{(1)} = \alpha^{(2)} = 0$, $\alpha^{(3)} > 0$ and so on. The ordering is evidently not unique, but for any definite choice positivity is completely defined, and induces an ordering of the roots, namely, $\alpha > \beta$ if $\alpha-\beta$ is positive. It is evident that for each ordering of the H_i the roots split evenly into positive and negative sets, and that there is a unique set of l linear independent smallest positive roots. *These l smallest positive roots are called the primitive roots.* It is obvious that all other positive roots are linear combinations of the primitive roots with positive coefficients. What is not so obvious, but can be proved by using the step operations of section 3.5 is that the coefficients are *integers* (see, for example, Helgason, 1978 or Wybourne, 1974). Thus one has the remarkable result that any root can be written in the form

$$\alpha = \pm \sum_{i=1}^{l} n_i \alpha_i, \tag{3.37}$$

where α_i are the l primitive roots and the n_i are non-negative integers. For example, with H_1 horizontal and H_2 vertical, the primitive roots for the $l = 2$

algebras are those labelled α_1 and α_2 in fig. 1. The derivation of (3.37) also shows that the root system is completely determined by the primitive roots (i.e. the possible integers n_i are fixed once the α_i are fixed) and thus the compact simple Lie algebras are completely characterized by the primitive roots. For this reason it is very important to know the structure of the primitive root system, and since its only invariant ingredients are the lengths and inner products of the roots, it is usually summarized as the matrix

$$C_{ij} = 2\langle \alpha_i, \alpha_j \rangle / \langle \alpha_j, \alpha_j \rangle, \quad i,j = 1, \dots, l, \qquad (3.38)$$

which is known as the Cartan matrix. (A symmetrical version of it, called the Coxeter matrix, is obtained by changing the denominator to $|\alpha_i||\alpha_j|$.) For the classical groups, the Cartan matrix has only diagonal or next-to-diagonal elements ($C_{ij} = 0$ unless $i = j, j \pm 1$), and a simple graphical way of displaying it for all simple groups (due to Coxeter and Dynkin) is the following: Let a small circle denote a primitive root and let three, two, one, zero lines between circles denotes angles $\frac{5\pi}{6}$, $\frac{3\pi}{4}$, $\frac{2\pi}{3}$ and $\frac{1}{2}\pi$ between the corresponding roots. Also, let an arrow indicate the direction from a longer to a shorter root. For example, for the weight diagrams of fig. 1, one would have the Cartan matrices and diagrams

$$SU(3): \begin{pmatrix} 2 & -1 \\ -1 & 2 \end{pmatrix} \approx \text{O——O}$$

$$SO(5): \begin{pmatrix} 2 & -1 \\ -2 & 2 \end{pmatrix} \approx \text{O⟹O}$$

$$G_2: \begin{pmatrix} 2 & -3 \\ -1 & 2 \end{pmatrix} \approx \text{O⟹O}$$

respectively. The primitive root systems for the simple compact algebras of rank l are then in one–one correspondence with the connected diagrams with l circles. The connected diagrams are called Dynkin diagrams, and the problem of classifying the simple compact Lie algebras reduces to the problem of classifying all such diagrams. This is not a difficult problem to solve, and the solution produces quickly the standard classification, which was due originally to Cartan. According to this classification the simple compact Lie algebras are as in table 1 (below), where the letters A to G refer to the (complexified) algebra. Thus there are four main classes of compact simple Lie algebras and they correspond to the four classes of classical groups. For these H_i are the diagonal and E_α the off-diagonal matrices. But in addition there is an exceptional class E_l with three Lie algebras, and two exceptional Lie algebras F_4 and G_2. The five exceptional

Table 1. *Classification of the compact simple Lie algebras.*

Class	Range of l	Order	Compact group	Dynkin diagram
A_l	$l = 1, 2, 3, \ldots$	$r = (l+1)^2 - 1$	$SU(l+1)$	o—o ··· o—o—o
B_l	$l = 1, 2, 3, \ldots$	$r = l(2l+1)$	$SO(2l+1)$	o—o ··· o—o⇒o
C_l	$l = 1, 2, 3, \ldots$	$r = l(2l-1)$	$Sp(2l)$	o—o ··· o—o⇐o
D_l	$l = 3, 4, 5, \ldots$	$r = l(2l-1)$	$SO(2l)$	o—o ··· o—o<
E_l	$l = 6, 7, 8$	$r = 78, 133, 248$	E_l	o—o—o ··· —o
F_4	$l = 4$	$r = 52$	F_4	o—o⇒o—o
G_2	$l = 2$	$r = 14$	G_2	o⇛o

algebras are connected with the algebraic numbers. For example, just as $SU(2)$ is the group of automorphisms of the quaternions (Pauli matrices), G_2 is the group of automorphisms of the octonions and F_4 the group of automorphisms of the Jordan algebra (hermitian 3×3 matrices with octonion entries). G_2 can also be defined as the group which leaves one direction in the eight-dimensional spinor representation of $SO(8)$ (chapter 4) invariant.

The algebras D_1 and D_2 do not occur on the list because $D_1 = SO(2)$ is abelian and $D_2 = D_1 + D_1$ is a direct sum, and there are a few isomorphisms among the lower rank algebras which correspond to homomorphisms among the classical groups, namely,

$$SO(3) = SU(2)/Z_2 = Sp(2)/Z_2, \quad (B_1 = A_1 = C_1)$$

$$SO(5) = Sp(4)/Z_2 \quad (B_2 = C_2) \tag{3.39}$$

$$SO(6) = SU(4)/Z_2 \quad (D_3 = A_3).$$

All the compact groups listed are simply connected, except the orthogonal groups, each of which has a double covering,

$$SO(n) = \widetilde{SO}(n)/Z_2.$$

It may also be useful to list the discrete centres. These are all of the form Z_p, where Z_p is the cyclic group of order p, and for the classical groups they are easily seen to be

$SO(2l+1)$: $Z = 1$ (centre is trivial).
$SO(2l)$, $Sp(2l)$: $Z = Z_2 (= \{1, -1\})$ where 1 is the unit $2l \times 2l$ matrix.
$SU(n)$: $Z = Z_n (= 1 \exp 2\pi i m/n)$ where $m = 1, \ldots, n$ and 1 is the unit $m \times m$ matrix.

For the exceptional groups they are (Tits, 1967)

$$Z(G_2) = Z(F_4) = Z(E_8) = \mathbb{1}, \quad Z(E_7) = Z_2, \quad Z(E_6) = Z_3.$$

Finally, although non-compact Lie groups are not in general use as gauge groups (with the exception of the Lorentz and Poincaré groups for gravitation) it might be mentioned for completeness that the algebras $A, \ldots, G$ above classify also the (real) non-compact semi-simple groups, as follows: Since for each semi-simple algebra the Cartan metric g is non-degenerate, it can be chosen as $g = \text{diag}(\mathbb{1}_p, -\mathbb{1}_q)$, and since its adjoint representation A is skew-symmetric with respect to g ($A^t G + G A = 0$) there exists a basis $A = K + M$ for A such that, symbolically,

$$[K, K] = K, \quad [K, M] = M, \quad [M, M] = -K, \tag{3.40}$$

where $K^t = -K$, $M^t = M$. The algebra K is actually the maximal compact subalgebra. The algebra A admits the isomorphism $K \to K$, $M \to -M$ and one sees that if one lets $M \to iM$ then $g \to \text{diag}(\mathbb{1}_p, \mathbb{1}_q)$, which is the metric for a compact algebra. Thus each non-compact semi-simple algebra can be made compact by the attachment of factors i to the appropriate generators M (a procedure that is known as the 'unitary trick' because it makes the adjoint representation of the group unitary). It is not difficult to see that each compact algebra has a finite, indeed a very limited, number of inequivalent non-compact partners, and hence that the classification of the compact algebras and the associated unitary tricks suffices to classify all the non-compact semi-simple algebras. A complete classification of this kind was first given by Cartan and has been summarized for physicists by Barut and Raczka (1965).

Exercises

3.1. Express the conditions (a), (b), ..., (f) of section 3.1 for subalgebras, etc. as conditions on the structure constants (e.g. $f_{ik}^\alpha = 0, \alpha \neq j$ where $i, j, k = 1, \ldots, p$ are the indices of a subalgebra).

3.2. Show that for $SU(n)$ the Cartan algebra can be chosen as the $n-1$ independent traceless diagonal matrices M, the positive roots as real upper-triangular matrices with one entry, and the primitive roots as the immediately off-diagonal roots $M_{ij} = \delta_{j, i+1}$. What are the corresponding results for $SO(2n)$ and $SO(2n+1)$?

3.3 Show from its action on the roots that the adjoint representation of the Weyl group for $SU(3)$ is the group of permutations of three objects, and find its action on the Cartan elements

$$T_3 = (2)^{-1} \text{diag}(1, -1, 0), \quad Y = (12)^{-\frac{1}{2}} \text{diag}(1, 1, -2).$$

4
Hermitian irreducible representations of compact simple Lie algebras

4.1 Weight diagrams, dominant and highest weights

It will be seen in the next chapter that the representation theory of the compact connected Lie groups depends largely on the hermitian irreducible representations (HIRs) of the Lie algebra, and that these are all finite dimensional. For this reason it is convenient to consider first the HIRs of the simple compact algebras.

The most convenient basis for the description of the HIRs is the Cartan basis described in chapter 2. The representative matrices in this basis are then $U(H_i)$, $U(E_\alpha)$ and $U(E_{-\alpha}) = U^+(E_\alpha)$, where $U(g)$ is the continuous unitary irreducible representation (CUIR) of the corresponding group, but for brevity the U will be dropped and the representative matrices will be referred to as H_i, E_α and $E_{-\alpha}$. Note that the H_i, $E_\alpha + E_{-\alpha}$ and $i(E_\alpha - E_{-\alpha})$ are then hermitian matrices.

The base vectors in the representation are chosen to be eigenvectors of the H_i, i.e.

$$H_i|m\rangle = m_i|m\rangle \quad \text{or} \quad \mathbf{H}|m\rangle = \mathbf{m}|m\rangle, \tag{4.1}$$

where the m_i are the simultaneous eigenvalues of the H_i and need not be simple, i.e. the eigenspaces $|m\rangle$ need not be one dimensional. From the Cartan commutation relations (3.28) one sees that the action of the remaining generators E_α on the eigenspaces $|m\rangle$ is

$$E_\alpha|m\rangle = |m+\alpha\rangle. \tag{4.2}$$

It follows that if the eigenvalues m_i are depicted as vectors in an l-dimensional Euclidean space they form a lattice with spacings given by the roots. The l-vectors m in such a lattice are called *weights* and the lattice itself is called a *weight diagram*. Some examples are given in fig. 2, the multiplicity of the weights being indicated by the number of circles around the weights.

The argument that was used in section 3.6 to show that the roots form

36

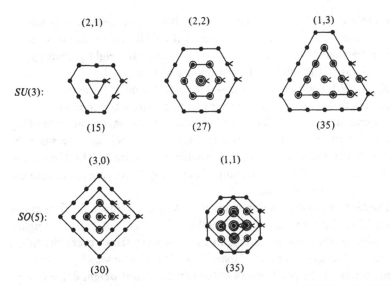

Fig. 2. Weight diagrams for some irreducible representations of $SU(3)$ and $SO(5)$. The Dynkin indices (λ, μ) are placed above the diagrams and the dimensions (d) below. The multiplicity of each weight is one plus the number of circles around it. In particular, the weights without circles are simple. Dominant weights are marked $<$.

(finite) 'strings' can be used to show that the weights also form finite strings,

$$m - s\alpha, m - (s-1)\alpha, ..., m - \alpha, m, m + \alpha, ..., m + (t-1)\alpha,$$

$$\frac{2(m, \alpha)}{(\alpha, \alpha)} = \text{integer}, \tag{4.3}$$

where α is any root. It follows in particular that the weight diagrams, like the root diagrams, are invariant with respect to the Weyl group, and it is this property and (4.3) that gives them the crystallographic structure seen in fig. 2. Note that the weights $\mathbf{m}$ lie on orbits of the Weyl group, and since the Weyl group preserves length, each orbit lies on a sphere. The weights can also be ordered according to the ordering introduced in section 3.7, namely, $m > m'$ if $m_1 > m'_1$ or $m_1 = m'_1$, $m_2 > m'_2$ and so on, and the most positive weight in a Weyl orbit, according to this ordering, is called the *dominant* weight $\mathbf{d}$ of the orbit (fig. 2). In an irreducible representation HIR there will, in general, be a number of Weyl orbits, each with a dominant weight, and the most positive of the dominant weights, and thus the most positive weight in the HIR, is called the *highest* weight $\mathbf{j}$ of the IR. From (4.2) it satisfies the condition

$$E_\alpha |j\rangle = 0, \quad \alpha > 0. \tag{4.4}$$

The highest weight **j** of an HIR is very important because it is always simple, and it can be used to characterize the HIR. To see that it is simple one notes that if $|j\rangle'$ is another state with highest weight **j**, then by the irreducibility and finite dimensionality, it must be of the form $E_\beta E_\gamma, ..., E_\delta |j\rangle$ where $\alpha + \beta + ... + \delta = 0$. But then the Cartan commutation relations can be used to move all the E_ϵ with $\epsilon > 0$ to the right until they annihilate $|j\rangle$ according to (4.4), until finally only a multiple of $|j\rangle$ remains. This shows that $|j\rangle'$ is a multiple of $|j\rangle$. Furthermore since the action of the E_β, $E_{-\beta}$, H_i on $|j\rangle$ is completely determined by the Cartan commutation relations on account of (4.4), any HIRs generated from the same $|j\rangle$ must be the same.

The highest weight is a dominant weight by definition, and while in any given HIR there are dominant weights which are not highest weights, nevertheless the converse is true in the following sense: every dominant weight is the highest weight of *some* HIR. This can be seen by noting that if one demands the condition (4.4) for any dominant weight **d**, the Cartan algebra can be realized on the space generated from $|d\rangle$ by the $E_{-\alpha}$ and can produce no higher weight. Thus the HIRs are in one–one correspondence with the dominant weights, a result that will be of great use later.

Finally it is worth noting that since the HIRs are characterized both by the conventional characters $\chi(k)$ and by the highest weights, there must be a relationship between these two. Let us first consider this relationship for the familiar case of the three-dimensional rotation group $SO(3)$. Since for this group the conjugation classes are defined by rotations ϕ about a fixed axis (section 1.1), one has

$$\chi(\phi) = \text{tr}\,(\exp i\phi\sigma_3) = \sum_{-j}^{j} \exp im\phi = \frac{\sin(j+\tfrac{1}{2})\phi}{\sin\tfrac{1}{2}\phi}, \qquad (4.5)$$

where ϕ is the third Euler angle, σ_3 the generator of rotations about the third axis, and m is the magnetic quantum number. Note that the last expression in (4.5) involves only the highest weight j.

For the general compact simple Lie algebra the situation is very similar. It can be shown that any element may be conjugated into a Cartan element, and hence the conjugation classes are defined by the Cartan algebra alone. For this algebra one has

$$\chi(\phi) = \text{tr}\,\exp i(\phi, H) = \sum_m \gamma(m)\,e^{i(m,\,\phi)} = \sum_w \epsilon(w)\,e^{i(w(j+\delta),\,\phi)} \Big/ \sum_w \epsilon(w)\,e^{i(w\delta,\,\phi)},$$

$$(4.6)$$

where $\gamma(m)$ is the multiplicity of the weight m, the sum is over all elements

w of the Weyl group W, $\epsilon(w) = (-1)^n$ where n is the number of Weyl reflexions in w, and δ is half the sum of the positive roots. The last term in (4.6) is known as Weyl's character formula and for its derivation the reader is referred to the references (e.g. Humphreys, 1972). Equation (4.5) is obviously a special case of (4.6) and in both formulae the relationship between the character and the highest weight is evident.

4.2 Multiplicities of internal weights

It has been seen that highest weights $\mathbf{j}$ of the HIRs of simple compact groups are simple. This is not true, in general, for the other weights in an HIR, though it may be true in special cases such as the symmetric tensor representations of $SU(n)$ (and therefore all HIRs of $SU(2)$). Even in the adjoint representations where the non-zero weights are simple (since they are roots) the zero weight has, by definition, multiplicity l. For most representations both the zero and non-zero weights m will have multiplicities $\gamma(m)$ greater than unity and two examples are shown in fig. 2. In the examples of this figure it is seen that the multiplicity tends to increase in the inward direction, and it turns out that this is a general feature – the multiplicity in any HIR is a monotonically decreasing function of distance from the centre. Note that the multiplicity is the same for all weights on the same Weyl orbit. In particular, the multiplicity is unity for all weights on the same orbit as the highest weight $\mathbf{j}$. Thus the statement that the highest weight is simple generalizes to the invariant (ordering-independent) statement that the outermost Weyl orbit is simple. This means, of course, that any weight on this orbit could be used as highest weight with an appropriate ordering.

Apart from some special results for low-dimensional HIRs there is no simple formula for the multiplicities. The best one can do is either to express them in terms of a partition function $P(x)$ (Kostant, 1959; Bincer and Schmidt, 1984) or to use a recursion formula due to Freudenthal (Humphries, 1972). Kostant's formula is

$$\gamma(m) = \sum_w \epsilon(w) P(w(\mathbf{j}+\delta) - (\mathbf{m}+\delta)), \tag{4.7}$$

where $\mathbf{j}$ is the highest weight, 2δ is the sum over the positive roots, the sum is over the Weyl group with elements w, $\epsilon(w) = \pm 1$ according as the number of reflexions in w is even or odd, and the partition function $P(x)$ is defined as

$$P(\mathbf{x}) = \textit{number of sets of non-negative integers } k_\alpha \textit{ such that } \sum k_\alpha\, \mathbf{a} = \mathbf{x}, \tag{4.8}$$

where α is any positive root. Kostant's formula is direct and is therefore most useful when the multiplicity of one particular weight m is required. Freudenthal's recursion formula is

$$[(j+\delta)^2 - (m+\delta)^2]\gamma(m) = 2 \sum_{\alpha>0} \sum_k (m+k\alpha, \alpha)\gamma(m+k\alpha), \quad k = 1, 2, \ldots$$

(4.9)

where the sum over k is, of course, finite, and one sees that each $\gamma(m)$ is expressed in terms of the multiplicities $\gamma(m+k\alpha)$, $r = 1$, 2, ... of the weights outside m. Freudenthal's formula is the more useful one when the multiplicities of all the weights are required. The derivation of these formulae is beyond the scope of this book and for more information the reader is referred to the references (e.g. Humphreys, 1972).

4.3   The dual of the root lattice. Primitive weights

The fact that for every weight m in a HIR the quantity $2(m, \alpha)/(\alpha, \alpha)$ is an integer means that the weights are in a certain sense dual to the roots, and this idea may be formalized and used to classify the HIRs as follows: Let a_i be the l primitive roots $i = 1, \ldots, l$ and ω_i the l linearly independent vectors which satisfy the condition

$$2(\omega_i, \alpha_j) = \delta_{ij}(\alpha_i, \alpha_i).$$

(4.10)

The l-vectors ω_i will be called primitive weights, and for the classical groups and G_2 they are given in table 2. Note from the table that the generic form of ω is e_1, e_1+e_2, $e_1+e_2+e_3$, ..., but that there are small modifications for each individual group. Thus for SU $(l+1)$ a term is subtracted because it is easier to work in an $(l+1)$-dimensional space, for $SO(2l)$ the base vector e occurs with both signs, and for all the orthogonal groups there are some factors $\frac{1}{2}$. In the next section it will be shown by construction that the l primitive weights ω are actually the highest weights of l HIRs. These l HIRs will be called the l primitive representations F_k of the algebra, $k = 1, \ldots, l$, and for the moment it will be simply assumed that they exist.

Since the primitive weights ω_i are linearly independent, any weight $\mathbf{m}$ can be expanded in terms of them,

$$\mathbf{m} = \sum_{i=1} \mu_i \omega_i,$$

(4.11)

and because $2(m, \alpha)/(\alpha, \alpha)$ must be an integer it is easy to see that the coefficients μ_i must be integers. In general these integers are not positive, but from the definition of the Weyl group and its action on the primitive

Table 2. The roots, the primitive roots, and the primitive weights are given in terms of orthonormal base vectors in root space (except for $SU(n)$ for which it is more convenient to regard the root space as the plane orthogonal to the vector $e_0 = (e_1+e_2+...+e_n)/n$ in an $n (=l+1)$-dimensional space). The situation for the other exceptional groups is similar to that for G_2, i.e. the roots are obtained by adding some special combinations of the e_k to the roots of a certain classical group (Wybourne, 1974).

	$SU(l+1)$	$Sp(2l)$	$SO(2l+1)$	$SO(2l)$	G_2
Roots	$e_i - e_j$ $(i,j=1,...,l+1)$	$e_i - e_j$ $\pm 2e_i$ $\pm(e_i+e_j)\,(i\neq j)$	$e_i - e_j$ $\pm e_i$ $\pm(e_i+e_j)\,(i\neq j)$	$e_i - e_j$ $\pm(e_i+e_j)\,(i\neq j)$	$e_i - e_j$ $\pm 2e_i \mp e_j \pm e_k$
Primitive roots	$e_i - e_{i+1}$	$e_i - e_{i+1}$ $2e_l$	$e_i - e_{i+1}$ e_l	$e_i - e_{i+1}$ $e_{l-1}+e_l$	$e_3 - e_2$ $e_1 + e_2 - 2e_3$
Primitive weights	$e_1 - e_0$ $e_1 + e_2 - 2e_0$ ... $e_1+...+e_k - ke_0$ ... $e_1+...+e_l - le_0$	e_1 $e_1 + e_2$ $e_1 + e_2 + e_3$ ... $e_1+...+e_l$	e_1 $e_1 + e_2$ ... $e_1+...+e_{l-1}$ $\frac{1}{2}(e_1+...+e_l)$	e_1 $e_1 + e_2$ ... $e_1+...e_{l-2}$ $\frac{1}{2}(e_1+...+e_{l-1}\pm e_l)$	$e_1 - e_2$ $\frac{1}{3}(2e_1 - e_2 - e_3)$

The transformations $\alpha_i \to \tilde{\alpha}_i = 2\alpha_i/\alpha_i^2$ are involutory and interchange the long and short roots modulo the scale. Thus they rescale the root-diagrams except for B_l and C_l, which they interchange. Through (4.10) they induce the transformations $\omega_i \to \tilde{\omega}_i = 2\omega_i/\alpha_i^2$ and the 'coweights' $\tilde{\omega}$ have the property that, if the α are scaled so that for $0 \leq t \leq 1$ the $\exp(it\alpha \cdot H)$ are (single) closed loops, then the $\exp(it\tilde{\omega}\cdot H)$ are geodesics from 1 to the centre Z. However, only for the coweights of $(\omega_k, k = 1,...,l)$ $(\omega_1), (\omega_k, \omega_{l-1}, \omega_l), (\omega_1, \omega_{l-1}, \omega_l), (\omega_1, \omega_2)$ and (ω_1) of $SU(n)$, $Sp(2n)$, $SO(2n+1)$, $SO(2n)$, E_6 and E_7 are the geodesics minimal (do not loop), and these minimal coweights are important in the theory of monopoles and of representations of Kac-Moody algebras.

roots it can be verified that the sign of any non-zero integer μ_i can be reversed by a Weyl reflexion. It follows that for the dominant weights **d** defined in the previous section the integers μ_i are positive,

$$\mathbf{d} = \sum_{i=1}^{l} \mu_i \omega_i, \quad \mu_i \geqslant 0. \tag{4.12}$$

In the same way it can be seen that, conversely, if $\mu_i \geqslant 0$ then **d** is dominant.

Since the HIRs are in one–one correspondence with the dominant weights **d**, it follows that to each HIR there corresponds a set of non-negative integers μ_i, which defines it. The converse is also true, i.e. to every set of l non-negative integers μ_i there corresponds an HIR, and this can be seen as follows: Let $\mathbf{j}(p)$, $p = 1, \ldots, n$ be the highest weights of any n HIRs. Then since multiplication in the group corresponds to addition in the Lie algebra, the highest weight in the product representation of all n HIRs will be $\mathbf{j} = \sum_p \mathbf{j}(p)$. Applying this result to the product

$$(F_1 \times F_1 \times \ldots \times F_1) \times (F_2 \times F_2 \times \ldots \times F_2) \times \ldots \times (F_l \times F_l \times \ldots \times F_l), \tag{4.13}$$
$$\underbrace{\hphantom{(F_1 \times F_1 \times \ldots \times F_1)}}_{\mu_1} \quad \underbrace{\hphantom{(F_2 \times F_2 \times \ldots \times F_2)}}_{\mu_2} \quad \underbrace{\hphantom{(F_l \times F_l \times \ldots \times F_l)}}_{\mu_l}$$

of the primitive representations, where each μ_i is any non-negative integer, and which is totally symmetric in each F_l, one sees that the highest weight in the product is just $\sum \mu_i \omega_i$. Since any non-negative integers can be used in (4.13) this establishes the result.

This is a rather remarkable result because it shows that the HIRs of the simple compact groups are completely characterized by the sets $(\mu_1, \ldots, \mu_l)$ of l non-negative integers (called Dynkin indices) and the l primitive HIRs. Furthermore, it gives an explicit construction of the HIRs as the leading irreducible representations in (4.13).

4.4 Fundamental tensor (single-valued) representations

Let us now consider the primitive representations. First, it must be noted that because the orthogonal groups $SO(n)$ are doubly connected, there exist representations (so called spinor representations) of the covering groups $\widetilde{SO}(n)$ which are not true, but rather two-valued, representations of the orthogonal groups themselves. These representations correspond to the factors $\frac{1}{2}$ in table 2 and their discussion is deferred to the next section. In this section only the tensor, or single-valued (though not necessarily faithful) representations will be considered.

The primitive tensor representations are constructed in essentially the same manner for each of the compact simple groups, namely, as the first

l totally antisymmetric products of the fundamental (defining) representation F,

$$F_1 = F, \quad F_2 = F \wedge F, \quad ... \quad F_s = \underbrace{F \wedge F \wedge ... \wedge F}_{s}, ... \quad s = 1, ..., l, \quad (4.14)$$

with some slight modifications for some of the groups. It is clear that the dimensions of the F_s are $\binom{d}{s}$ where $d = \dim F$, and that the highest weight is $\sum_{k=1}^{s} m_k$ where $m_1, m_2, ...$ are the weights of F in order of positivity. What has to be verified is that these weights are just the fundamental weights ω_s of the previous section. It is convenient to consider each Cartan class separately:

$SU(l+1)$: The weights of F are $m_i = e_i - e_0$, so the agreement with table 1 is obvious. Furthermore, the F_r are found to be irreducible and $F_r = F^*_{l+1-r}$. Thus (4.14) requires no modification. For example for $SU(3)$, $SU(4)$ and $SU(5)$ the primitive representations are $(3, 3^*)$, $(4, 6, 4^*)$ and $(5, 10, 10^*, 5^*)$ respectively.

$SO(2l+1)$: The weights of F are $\pm e_i$, 0 and hence the highest weights of the F_r are $e_1, e_1 + e_2, ...$ in agreement with table 1 (modulo the $\frac{1}{2}$). Furthermore, F_r are irreducible. The only new feature is that the F_{l+1-r} (which are actually equivalent to the F_r) are not included. Thus the dimension increases up to $r = l$. For example, for $SO(7)$ the dimensions are 7, 21 and 35.

$SO(2l)$: The weights of F are $\pm e_i$ so again there is agreement with table 1 (modulo the $\frac{1}{2}$). All the F_r are irreducible except F_l which splits into two representations $F_l^\pm$ (on the self-dual and anti-self-dual parts of its representation space). For example, for $SO(6)$ the dimensions of F_1, F_2, $F_3^\pm$ are 6, 15 and $10^\pm$. For odd l the $F_l^\pm$ are actually complex.

$Sp(2l)$: The weights of F are $\pm e_i$ so the situation is analogous to that for $SO(2l)$. The new features are that F_l does not split and the F_r for $2 \leqslant r \leqslant l$ are reducible. They are made irreducible by removing the symplectic traces. For example F_2 acts on the space $u_s v_t - u_t v_s - J_{st}(u, Jv)/2l$, where F acts on u and v, and J is the symplectic metric.

G_2: In this case F is seven-dimensional (see next section) and $F \wedge F = 14 + 7$. So $F_2 = 14$, which is the adjoint representation. For the other exceptional groups see the references (e.g. Tits, 1967).

By combining the results of this and the previous sections one sees that all the tensor HIRs of all the simple compact Lie groups may be constructed by the following two-step process:

(a)   Form the first l totally *anti*symmetric products of the fundamental representation and extract the leading HIRs (if necessary). These are the l primitive representations F_s.

(b)   Form the totally *symmetric* products $F_1^{\mu_1} \times F_2^{\mu_2} \times \ldots \times F_l^{\mu_l}$ and extract the leading IRs. These are the HIRs with Dynkin indices $(\mu_1, \ldots, \mu_l)$.

This procedure produces all the tensorial HIRs of all the simple compact algebras.

Let us consider some examples. For $SU(2)$ the bases of all the IRs are products $u_s u_t \ldots u_q$ where $u_r, r = 1, 2$ is a basis for the fundamental representation. For $SU(3)$ the two primitive representations are the 3 and $3 \wedge 3 = 3^*$ and hence every HIR is the leading HIR in a product $(3)^\lambda \times (3^*)^\mu$, where λ and μ are the Dynkin indices. In tensor notation the base vectors for the HIR (λ, μ) may be written as

$$T_{rs\ldots t}^{ij\ldots k} - \delta_r^i \, T_{qs\ldots t}^{qj\ldots k}, \tag{4.15}$$

where there are λ upper and μ lower indices, and T is completely symmetric in each set. Note that a contraction between an upper and lower index is $SU(3)$-invariant, while a contraction between two lower or two upper indices is only invariant with respect to the real subgroup $SO(3)$.

For $SU(4)$ there are three primitive representations, 4, 6 and 4^*, and the 6 is actually the fundamental tensor representation of $SO(6)$, which has the same Lie algebra as $SU(4)$. For $SU(5)$ the primitive representations are 5, 10, 10^*, 5^* and it is interesting to find that these are just the representations that are needed to classify the elementary fermions (quarks, leptons), and their antiparticles, in $SU(5)$ grand unification theory. Finally the exceptional group E_8 has the interesting property that it is the only compact simple group for which the fundamental representation and the adjoint representation are the same. For this reason it is of particular interest for supersymmetric theories (Fayet and Ferrarra, 1977; Ferrarra, 1984; Bagger and Wess, 1983).

Although the construction given above for the HIRs of the compact simple algebras is very convenient from the point of view of classification, it is not always the most convenient for the actual construction. For example, it is often convenient to construct the HIRs directly from the fundamental representation F by means of a tensor T_Y, which is neither completely symmetric nor antisymmetric but has the mixed symmetry of a Young tableau Y. Such tensors T_Y carry irreducible representations of the general linear unimodular $SL(n, c)$ and hence carry representations of the classical groups which are irreducible up to the contractions that are permitted by the definitions of these groups. Details of Young tableau may be found readily, e.g. in Bacry (1977), Gourdin (1982).

4.5   Fundamental spinor representations

As already mentioned, the primitive tensor representations do not form a complete set for the orthogonal groups $SO(n)$, because they do not include the spinor representations (faithful representations of $\widetilde{SO}(n)$ which are two-valued representations of $SO(n)$). The best known example of a spinor representation is the two-valued, two-dimensional representation of $SO(3)$ provided by the fundamental representation of $SU(2)$. Other well-known examples are the two-valued, four-dimensional representations of $SO(5)$ and $SO(6)$ provided by the fundamental representations of $Sp(4)$ and $SU(4)$. Since the 4 and 4* of $SU(4)$ are inequivalent one sees that $SO(6)$ has actually two inequivalent spinor representations.

These three examples are well known because the spinor representations happen to be tensor representations of other classical groups. In general this is not the case, but nevertheless the pattern is similiar. In particular, there is one fundamental spinor representation Δ for $SO(2l+1)$ and two (equidimensional) fundamental spinor representations $\Delta^{\pm}$ for $SO(2l)$. The representations Δ and $\Delta^{\pm}$ are two-valued, and constitute all the primitive spinor representations (i.e. for spinor representations, primitive = fundamental).

The construction (Boerner, 1970) of the fundamental spinor representations $\Delta, \Delta^{\pm}$ is a straightforward generalization of the well-known Dirac construction for the Lorentz group. Let

$$\{\gamma_\mu, \gamma_\nu\} = 2\delta_{\mu\nu}, \quad \gamma = (i)^l \gamma_1 \gamma_2 \cdots \gamma_{2l}, \tag{4.16}$$

be the (2^l-dimensional) representation of the $2l+1$ hermitian Dirac matrices γ_μ, γ. Then the generators of Δ and $\Delta^{\pm}$ are

$$\Delta(SO(2l+1)): \quad \tfrac{1}{4}[\gamma_\mu, \gamma_\nu], \quad \tfrac{1}{4}[\gamma_\mu, \gamma], \tag{4.17}$$

$$\Delta^{\pm}(SO(2l)): \quad \tfrac{1}{8}(1 \pm \gamma)[\gamma_\mu, \gamma_\nu], \tag{4.18}$$

respectively. Note that γ is $SO(2l)$-invariant and that the dimensions of Δ and $\Delta^{\pm}$ are 2^l and 2^{l-1} respectively. Note also that by adding γ_μ and γ to the generators of $SO(2l+1)$ one obtains the generators for one of the spinor representations of $SO(2l+2)$. A convenient Cartan basis is

$$H_i = \tfrac{1}{4}[\gamma_i, \gamma_{i+l}] \text{ for } SO(2l+1) \quad \text{and} \quad H_i = \tfrac{1}{8}(1 \pm \gamma)[\gamma_i, \gamma_{i+l}] \text{ for } SO(2l). \tag{4.19}$$

An explicit representation of the γs in terms of Pauli matrices is given in exercise 4.4, and from this representation one sees that the highest weights of Δ and $\Delta^{\pm}$ are $\tfrac{1}{2}(e_1 + e_2 + \ldots + e_l)$ and $\tfrac{1}{2}(e_1 + e_2 + \ldots + e_{l-1} \pm e_l)$ respectively. Thus they are exactly the weights needed to complete the primitive

The primitive tensor representations may be recovered from the primitive spinor ones by constructing bilinears of the form $\psi^\dagger \Gamma_A \psi$ where ψ are the spinors and Γ_A are monomials in the γ_μ. In this way one finds that

$$\Delta \times \Delta = F_l + F_{l-1} + \ldots + F_2 + F_1 + 1, \tag{4.20}$$

and, on decomposition into the symmetric and antisymmetric parts (with $q = 0, 1$),

$$(\Delta \times \Delta)_S = F_l + F_{l-2} + \ldots + F_q, \tag{4.21}$$

$$(\Delta \times \Delta)_{AS} = F_{l-1} + F_{l-3} + \ldots + F_q. \tag{4.22}$$

Similarly for Δ^+ and Δ^- one obtains (again with $q = 0, 1$)

$$\Delta^+ \times \Delta^- = F_{l-1} + F_{l-3} + \ldots + F_q, \tag{4.23}$$

$$\Delta^\pm \times \Delta^\pm = F_l^\pm + F_{l-2} + \ldots + F_q, \tag{4.24}$$

and, on decomposition of (4.24) into symmetric and antisymmetric parts (with $0 \leqslant q \leqslant 3$)

$$(\Delta^\pm \times \Delta^\pm)_S = F_l^\pm + F_{l-4} + \ldots + F_q, \tag{4.25}$$

$$(\Delta^\pm \times \Delta^\pm)_{AS} = F_{l-2} + F_{l-6} + \ldots + F_q. \tag{4.26}$$

When the spinor representations are taken into account the rules of section 4.4 for constructing the general HIRs of the compact simple algebras are modified as follows:

(a) Replace the primitive representation F_l by Δ for $SO(2l+1)$, and the primitive representations F_{l-1} and F_l ($= F_l^+ + F_l^-$) by $\Delta^\pm$ for $SO(2l)$.

(b) Form the totally symmetric products

$$(F_1)^{\mu_1} \times (F_2)^{\mu_2} \times \ldots \times (F_{l-1})^{\mu_{l-1}} \times \Delta^\mu \quad \text{for} \quad SO(2l+1),$$

and

$$(F_1)^{\mu_1} \times (F_2)^{\mu_2} \times \ldots \times (F_{l-2})^{\mu_{l-2}} \times (\Delta^+)^{\mu_+} \times (\Delta^-)^{\mu_-} \quad \text{for} \quad SO(2l),$$

and extract the leading HIRs.

The new Dynkin indices are evidently $(\mu_1, \ldots, \mu_{l-1}, \mu)$ and $(\mu_1, \ldots, \mu_{l-2}, \mu_+, \mu_-)$.

Let us consider some low-rank examples. For $SO(3)$ the fundamental tensor representation is the three-dimensional vector representation, but if the spinor representations are admitted it is replaced by the two-dimensional spinor representation Δ, and the construction of all other IRs is the same as for $SU(2)$ in section 3.5. For $SO(5)$ the primitive tensor representations are the 5 and 10 (i.e. the five- and ten-dimensional representations) but if spinor representations are admitted the 10 is replaced by the fundamental spinor representation Δ, which is four-dimensional, and the subsequent construction is the same as for $Sp(4)$. For $SO(7)$ the primitive tensor representations are the 7, 21 and 35 but if spinor representations are admitted the 35 is replaced by Δ which is eight-dimensional. For $SO(6)$ the primitive tensor representations are the 6, 15 and $20 = 10^+ + 10^-$ but if spinor representations are admitted the 15 and $10^+ + 10^-$ are replaced by the spinor representations $\Delta^\pm = 4, 4^*$ and the subsequent construction is the same as for $SU(4)$. Finally for $SO(8)$ the primitive tensor representations are the 8, 28, 56 and $70 = 35^+ + 35^-$, but if the spinor representations are admitted the 56 and 70 are replaced by $\Delta^\pm$ which are eight-dimensional. Note that $SO(8)$ then has three eight-dimensional primitive representations.

In conclusion it might be worth remarking that the exceptional group G_2 may be defined as the group which leaves a one-dimensional subspace of the eight-dimensional space of the $\Delta^\pm$ representation of $SO(8)$ invariant. This is why the lowest dimensional representation of G_2 is seven-dimensional.

4.6 Diagrammatic representation of the Dynkin indices

A useful graphical construction of the Dynkin indices of the compact simple groups may be obtained by inserting the indices in the circles of the Dynkin diagrams. Thus for $SU(n)$ one may write

$$\boxed{\mu_1} - \boxed{\mu_2} - \; \cdots \; - \boxed{\mu_{l-1}} - \boxed{\mu_l}$$

For example the 3, 3*, adjoint, 27-dimensional and rank n symmetric tensor representations of $SU(3)$ would be

$$\boxed{1} - \boxed{0} \quad \boxed{0} - \boxed{1} \quad \boxed{1} - \boxed{1} \quad \boxed{2} - \boxed{2}$$

$$\boxed{n} - \boxed{0} \quad \boxed{0} - \boxed{n}$$

respectively. For $SO(2l+1)$ the spinor index μ is placed in the exceptional circle which is attached by two lines, i.e. one has

$$\boxed{\mu_1} - \boxed{\mu_2} - \; \cdots \; - \boxed{\mu_{l-1}} = \boxed{\mu}$$

For example, the three primitive representations of $SO(7)$ (and F_3) are

$$7 = \boxed{1} - \boxed{0} = \boxed{0} \quad 21 = \boxed{0} - \boxed{1} = \boxed{0}$$

$$\Delta = \boxed{0} - \boxed{0} = \boxed{1} \quad F_3 = \boxed{0} - \boxed{0} = \boxed{2}$$

For $SO(2l)$ the spinor indices $\mu_\pm$ are placed in the exceptional circles which are not colinear with the others, i.e.

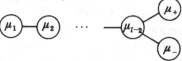

For example the four primitive representations of $SO(8)$ are

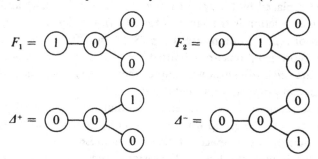

while the 'tensorial primitive' representations F_3 and $F_4^\pm$ are

$$F_3 = \text{(diagram)}$$

$$F_4^+ = \text{(diagram)}$$

$$F_4^- = \text{(diagram)}$$

Exercises

4.1. The primitive roots and their duals, the primitive weights, are given in table 2. In the case of the classical groups, verify that they do indeed satisfy the dual relationship (4.10).

4.2. Show that none of the weights in the primitive or symmetric tensor $(n, 0, 0, ..., 0)$ and $(0, 0, ..., n)$ representations of $SU(n)$ have multiplicity greater than one.

4.3. For the group $SO(10)$, verify that the dimensions of the primitive tensor, adjoint, symmetric tensor and fundamental spinor representations are (10, 45, 120, 210, $126^\pm$), 45, 54 and $16^\pm$ and write out the decompositions of $16^\pm \times 16^\pm$ and $16^+ \times 16^-$ in terms of the primitive tensor representations.

4.4. Show that a representation of the Dirac algebra (4.16) is given by

$$\gamma_i = \sigma_3 \times \sigma_3 \times ... \times \sigma_3 \times \sigma_1 \times 1 \times ... \times 1$$
$$\gamma_{i+p} = \sigma_3 \times \sigma_3 \times ... \times \sigma_3 \times \sigma_2 \times 1 \times ... \times 1$$
$$\gamma = \sigma_3 \times \sigma_3 \times ... \times \sigma_3$$

where the σs are the Pauli matrices, each product has p factors, and σ_1, σ_2 are in the ith position.

5
Continuous unitary irreducible representations (CUIRs) of compact Lie groups

5.1 Representations of Lie groups

In chapter 1, Lie groups were treated as abstract entities in order to clarify their structure. Their historical and practical importance, however, lies in the fact that they occur naturally as transformation groups, i.e. as representations – rotations of Euclidean 3-space, Lorentz transformations, unitary transformations on Hilbert space, and so on. Generally the transformations are of some finite-dimensional manifold M, with local coordinates x^α say, or of some differentiable functions $\psi(x)$ on such a manifold. Thus

$$x^\alpha \xrightarrow{g(a)} f^\alpha(x,a), \quad \psi(x) \xrightarrow{g^{-1}(a)} \psi(f(x,a)), \tag{5.1}$$

where

$$f(x, \phi(a,b)) = f(f(x,a)b), \quad f(x,0) = x,$$

the $\phi(a,b)$ being the structure functions of the group. The first equality in (5.1) is the statement that the transformations $f(x,a)$ satisfy the group multiplication, i.e. form a homomorphic image, or *representation*, of the group. Note that the representation need not be faithful or irreducible or (at this stage) even linear, and that $f(x,a)$ reduces to $\phi(b,a)$ in the special case $M = G$.

As in the case of the $\phi(a, b)$, for analytic $f(x, a)$ the equality in (5.1) can be written in differential form, namely,

$$\frac{\partial f(x,b)}{\partial b} = F(f)v(b), \tag{5.2}$$

where

$$F(x) = \left(\frac{\partial f(x,a)}{\partial a}\right)_{a=0},$$

and $v(b)$ is the matrix defined in (2.4). It is easy to see that when the $v(b)$ satisfy the integrability conditions (2.7), the integrability conditions for the Fs are similar to those for the us in (2.9), namely,

$$[\mathscr{D}_s, \mathscr{D}_q] = f^t_{sq}\mathscr{D}_t, \quad \mathscr{D}_s = F^\alpha_s(x)\frac{\partial}{\partial x^\alpha}, \tag{5.3}$$

49

where f_{sq}^t are the structure constants of the group. Thus for every representation $f(x, a)$ the derivations $F(x)$ satisfy (5.3) and, conversely, every set of functions $F(x)$ satisfying (5.3) generate a local group transformation according to (5.2). In analogy to (2.24) the local representation can be obtained by exponentiation

$$\psi(f(x, a)) = e^{a \cdot \mathscr{D}} \psi(x), \tag{5.4}$$

for any analytic functions $\psi(x)$ on M.

Although nonlinear group representations have played some role in physics (they provide a fast method of deriving tree-graph results in current algebra (Gasiorowicz and Geffen, 1969) for example) most of the representations used are linear, and, since this is particularly true of gauge theories, only linear representations will be considered from now on. For the linear representations the functions $f(x, a)$ reduce to

$$f^\alpha(x, a) = \mathscr{D}_\beta^\alpha(a) \, x^\beta, \tag{5.5}$$

and the $\mathscr{D}(a)$ are matrices (more generally, bounded operators on a Hilbert space) which are independent of x, satisfy the group multiplication law, and are assumed to be continuous in the group topology,

$$\mathscr{D}(a)\mathscr{D}(b) = \mathscr{D}(\phi(a, b)), \quad \mathscr{D}(a) \to \mathscr{D}(b) \quad \text{as } a \to b. \tag{5.6}$$

5.2 Linear representations of compact Lie groups

Since the Lie groups which play a role in gauge theories (apart from gravitation) are compact, the linear representations of the compact groups are the ones of greatest interest. On account of the finiteness of the Haar measure, the representations (more precisely the linear, continuous, bounded representations) of the compact groups resemble those of the finite groups.

In particular they have the following properties:

(i) They are equivalent to unitary representations. To see this let $\mathscr{D}(g)$ be a continuous bounded representation on a space with inner product (f, h), and define a new inner product

$$\langle f, h \rangle = \int d\mu(g) \, (\mathscr{D}(g)f, \mathscr{D}(g)h), \tag{5.7}$$

by averaging with respect to the Haar measure. Then $\mathscr{D}(g)$ is seen to be unitary with respect to $\langle f, h \rangle$. From now on $\mathscr{D}(g)$ will be written as $U(g)$ to emphasize that it is unitary.

(ii) The continuous unitary irreducible representations (CUIRs) are finite dimensional. To see this, let $\mathscr{H}$ be the representation space, f be any vector in it, and consider the bounded operator Ω defined as

$$\Omega h = \int d\mu(g)\,(f, U(g)\,h)\,U^{\dagger}(g)f, \qquad (5.8)$$

where $h \in \mathscr{H}$. This operator is easily seen to be group invariant and hence, by Schurs lemma, it is a multiple of the unit operator

$$U^{\dagger}(g)\,\Omega\, U(g) = \Omega \Rightarrow \Omega = \omega \mathbf{1}, \quad 0 < \omega < \infty. \qquad (5.9)$$

Then by taking the trace and using the unitarity one has

$$\dim U(g) = \operatorname{tr}(\Omega/\omega) = \omega^{-1}(f,f)\int d\mu(g) < \infty, \qquad (5.10)$$

as required.

(iii) Inequivalent CUIRs are orthogonal:

$$\int d\mu(g)\,U_{rs}^{\dagger}(g)\,V_{ij}(g) = 0 \quad (U \not\approx V). \qquad (5.11)$$

This follows from the finite dimensionality and Schurs lemma just as for finite groups. In particular, by taking the trace of U and V one obtains the orthogonality relation for the characters,

$$\int d\mu(k)\,\chi_u^*(k)\,\chi_v(k) = 0 \quad (U \not\approx V), \qquad (5.12)$$

where the measure $d\mu(k)$ over the conjugacy classes is that induced by $d\mu(g)$.

(iv) The CUIRs are complete, in the sense that any square-integrable function over the group can be expanded in terms of them,

$$f(g) = \sum_{\lambda, ik} C_{ik}^{\lambda}\, U_{ik}^{\lambda}(g), \qquad (5.13)$$

where

$$\int d\mu(g)\,|f(g)|^2 = \sum_{\lambda, ik} |C_{ik}^{\lambda}|^2,$$

where λ labels the CUIRs $U(g)$ and takes only discrete values. This result is known as the Peter–Weyl theorem and may be regarded as the generalization of the Fourier transform for the circle group and of Burnside's completeness theorem for finite groups. In fact the only change from the case of finite groups is that the discrete range of λ is not finite. The Peter–Weyl theory is discussed in detail in Wawrzynczyk (1984).

Since compact Lie groups are direct products of simple groups, modulo the centre (which may contain $U(1)$ and discrete groups) their study reduces essentially to that of simple compact Lie groups. For the latter the operators $U(g)$ are finite $(n \times n)$ matrices and have all the analyticity prop.:rties of the group discussed in chapter 1. In particular they can be differentiated to yield, using (2.17),

$$\frac{\partial U(g(a))}{\partial a} = U(a)\, U(v(a))\, \sigma, \qquad (5.14)$$

where

$$\sigma = \left(\frac{\partial U(g)}{\partial a}\right)_{a-0},$$

and exponentiated to yield a representation of the local group

$$U(a) = \exp a \cdot \sigma, \quad a \cdot \sigma = a^r \sigma_r. \qquad (5.15)$$

Here the σs are r $(n \times n)$ constant matrices and are related to the infinitesimal generators as follows,

$$I_k = (\sigma_k)^\alpha_\beta \, x^\beta \frac{\partial}{\partial x^\alpha} \qquad (5.16)$$

where $[I_s, I_t] = f^u_{st} I_u$. From (5.16) one sees that the σ_k satisfy the same commutation relations as the I_k and since $U(a)$ is unitary they are skew-hermitian. Choosing a basis so that they are trace-orthonormal, one then has

$$[\sigma_s, \sigma_t] = f^u_{st} \sigma_u, \quad \sigma^\dagger_s = -\sigma_s, \quad \mathrm{tr}\, \sigma_s \sigma_t = -2\delta_{st} I, \qquad (5.17)$$

where I is a constant for each irreducible part of the representation. Equation (5.15) shows that for the CUIRs of the simple compact groups one may pass freely from the algebra to the local group. In fact, since for finite dimensional σ, the $\exp a \cdot \sigma$ are not only real analytic but entire, one may continue the as indefinitely and so pass to a representation of the global group. This representation is then a single-valued, but not necessarily faithful, representation of the covering group $\tilde{G}$, and a faithful, but not necessarily single-valued, representation of the adjoint group $G = \tilde{G}/Z_0$, where Z_0 is the centre of $\tilde{G}$. Only for one of the locally isomorphic groups G_t where $G \leqslant G_t \leqslant \tilde{G}$, will the representation be both faithful and single-valued, and this group G_t will be called the 'true' group of the representation.

5.3 Global properties of the representations: true groups

From the results of the previous section it is clear that the natural way to treat the continuous unitary irreducible representations (CUIRs) of the compact groups is to exponentiate the hermitian irreducible representations (HIRs) of the simple subgroups, and then determine the global properties, in particular to determine the 'true' group $G_t = \tilde{G}/Z_t$. This problem will not be considered in its full generality here, but because of the great relevance for unified and grand unified gauge theories the true groups will be found for the simple compact Lie groups and for a class of unitary Lie groups with a $U(1)$ group in the centre.

For the simple groups it is convenient to consider each Cartan class separately. For the classes with no centre, i.e. $SO(2l+1)$, G_2, F_4, E_8, there is obviously no problem, and for the classes with centre Z_2, i.e. $G = SO(2l)$, $Sp(l)$, E_7, the true group is determined by the parity of the tensor representations, i.e. tensors of odd and even rank (with respect to the fundamental representation) exponentiate to CUIRs of G and G/Z_2 respectively. Similarly, the class $\widetilde{SO}(2l+1)$ has centre Z_2 and the odd- and even-order spinor representations exponentiate to CUIRs of $\widetilde{SO}(2l+1)$ and $\widetilde{SO}(2l+1)/Z_2 = SO(2l+1)$ respectively. Finally for E_6 with centre Z_3, the true group is E_6 if the rank of the tensor is 1 or 2 (mod 3) and E_6/Z_3 if it is 0 (mod 3). Thus the only complicated cases are $SU(n)$ which has centre Z_n and $\widetilde{SO}(2l)$ which has centre Z_4 or $Z_2 \times Z_2$ according as l is odd or even.

For $SU(n)$ it is convenient to introduce a (rank) index

$$t = \sum^{l} s\mu_s \quad (\bmod\, n) \tag{5.18}$$

where μ_s are the Dynkin indices, n is the order of the centre Z_n, and t modulo n is to be taken in the range $(1, 2, ..., n)$ and not $(0, 1, ..., n-1)$. Thus for $SU(2)$, $t = 2j\,(\bmod\, 2)$, where j is the total spin, and for $SU(3)$, t is the conventional triality $t = \lambda + 2\mu\,(\bmod\, 3)$. The index t characterizes the rank of the representation (modulo n) because the fundamental representation occurs s times in F_s, and F_s occurs μ_s times in $(\mu_1, ..., \mu_n)$. The problem is to characterize the true groups $\tilde{G}/Z$ in terms of t. For this purpose let ω be the generating element of Z_n ($\omega^n = 1$). In any CUIR of index t, ω is represented by ω^t. Now let f be the highest common factor (h.c.f.) of t and n ($t = ft_0$, $n = fn_0$, where t_0 and n_0 are relatively prime). Then $(\omega^t)^{n_0} = (\omega^n)^{t_0} = 1$ and n_0 is the smallest integer for which this is true. It follows that the true central group is $Z_{n_0} = Z_n/Z_f$ and hence that the true group is

$$G = \tilde{G}/Z_f \tag{5.19}$$

where $f = $ h.c.f. of (t, n). Let us consider some examples. For $SU(3)$, $t = 1$, 2, 3, hence $f = (1, 1, 3)$ and thus the true group is $SU(3)$ for $t = 1$, 2 and

$SU(3)/Z_3$ for $t = 3$ (e.g. 8, 10, 10*, 27). For $SU(4)$, $t = 1, 2, 3, 4$, hence $f = (1, 2, 1, 4)$ and thus the true group is $SU(4)$ for $t = 1, 3$, $SU(4)/Z_2$ for $t = 2$ and $SU(4)/Z_4$ for $t = 4$. For $SU(6)$, $t = 1, 2, 3, 4, 5, 6$, hence $f = 1, 2, 3, 2, 1, 6$, and the true group is $SU(6)$ for $t = 1, 5$, $SU(6)/Z_2$ for $t = 2, 4$, $SU(6)/Z_3$ for $t = 3$ and $SU(6)/Z_6$ for $t = 6$. In general, for n prime, $f = n$ for $t = n$ and $f = 1$ otherwise, so the true groups are $SU(n)/Z_n$ and $SU(n)$ respectively. For the adjoint representation $t = n$, so the true group is $SU(n)/Z_n$. This can also be seen by noting that the adjoint representation is real whereas Z_n is complex.

For $\widetilde{SO}(2l)$ it is necessary to consider even and odd l separately. For odd l, i.e. $\widetilde{SO}(4n+2)$ (prototype $\widetilde{SO}(6)$) the centre is Z_4 and a convenient rank index is
$$t = \delta_+ + 2\tau + 3\delta_- \quad (\text{mod } 4), \qquad (5.20)$$

where τ is a rank index for tensors alone ($= 0, 1$ for even, odd), and t distinguishes four types of representations, with true groups as shown:

$$\left.\begin{array}{l} t = 1: \Delta^+\text{-like representations} \\ t = 3: \Delta^-\text{-like representations} \end{array}\right\} : \widetilde{SO}(4n+2)$$

$$t = 2: \text{odd-rank tensors}: SO(4n+2)$$

$$t = 0: \text{even-rank tensors}: SO(4n+2)/Z_2.$$

Note that the $\widetilde{SO}(6)$ index agrees with that of $SU(4)$ in (5.18), as it should, since $SU(4)$ and $SO(6)$ are locally isomorphic.

For even l, i.e. $\widetilde{SO}(4n)$ (prototype $\widetilde{SO}((8))$), the centre is $Z_2 \times Z_2$ and there are two independent rank indices, namely,

$$t_\pm = \mu_\pm + \tau \quad (\text{mod } 2) \qquad (5.21)$$

where τ is a tensor index ($= 0, 1$ for even, odd) and these, also, distinguish four classes of representations, with corresponding true groups, namely

$$\left.\begin{array}{l} t_+ = 1, t_- = 0: \Delta^+\text{-like spinors} \\ t_+ = 0, t_- = 1: \Delta^-\text{-like spinors} \end{array}\right\} : \widetilde{SO}(4n)$$

$$t_+ = t_- = 1: \text{odd-rank tensors}: SO(4n)$$

$$t_+ = t_- = 0: \text{even-rank tensors}: SO(4n).$$

Note that the indices $t_\pm$ can be used even for the non-simple case $\widetilde{SO}(4n) = SU(2) \times SU(2)$, in which case they are just $t_\pm = 2j_\pm$, where $j_\pm$ are the total spins of the two $SU(2)$ subgroups. Thus, as far as the centre Z_4 is concerned, the CUIRs of $\widetilde{SO}(2n)$ behave like the well-known $D(j_+, j_-)$

representations of $\widetilde{SO}(4)$. This correspondence provides an easy method for recalling the $\widetilde{SO}(4n)$ classification.

As well as simple groups, groups with a one-dimensional continuous centre $U(1)$ play an important role in particle physics. The most important of these at present are the groups with the same algebras as $SU(2) \times U(1)$, $SU(3) \times U(1)$ and $SU(3) \times SU(2) \times U(1)$. Hence it is worthwhile to consider the true groups for the algebras $SU(p) \times U(1)$ and $SU(p) \times SU(q) \times U(1)$ where p and q are prime $(p \neq q)$. It will be recalled that $U(1)$ is the group $\exp i\phi$, $0 \leqslant \phi \leqslant 2\pi$. Thus its CUIRs are $\exp(im\phi)$, $m = 0, \pm 1, \pm 2, ...$, and the integer m is both the Dynkin index and the rank index of the CUIR.

For $SU(p) \times U(1)$, p prime, there are two global groups, namely, $SU(p) \times U(1)$ and the quotient group $SU(p) \times U(1)/Z_p = U(p)$. The latter group is obtained by identifying the elements of Z_p in $SU(n)$ and $U(1)$ respectively. The best-known example is the identification of the elements $\text{diag}(-1, -1)$ and (-1) in $SU(2) \times U(1)$, and the generalization is to set

$$\omega = \omega' \qquad (5.22)$$

where $\omega \in Z_p(SU(p))$, $\omega' \in U(1)$ $(\omega^p = 1)$. Note that $U(p)$ occurs naturally in the direct-product, p-dimensional representation $SU(p) \times U(1)$, where $SU(p)$ and $U(1)$ act on the same vectors, whereas $SU(p) \times U(1)$ occurs naturally in the direct sum, $(p+1)$-dimensional representation $SU(p) \times 1 + 1 \times U(1)$.

In order to decide which of the groups $SU(p) \times U(1)$ is the true group for a CUIR (μ, m) one notes that ω and ω' are represented in this CUIR by $(\omega)^t$ and $(\omega')^m$ respectively, and hence the identification of elements $\omega = \omega'$ will be maintained if, and only if,

$$t = m \quad (\text{mod}\, p). \qquad (5.23)$$

Thus the true group for (μ, m) is $U(p)$ if (5.23) is satisfied, and $SU(p) \times U(1)$ otherwise. Note that the defining representations just discussed are $(t, m) = (1, 1)$ and $(1, 0) + (0, 1)$ respectively and thus are special cases of this rule.

For $SU(p) \times SU(q) \times U(1)$, where p and q are unequal and prime, there are four global groups, namely $G = SU(p) \times SU(q) \times U(1)$ and the quotients

$$U(p) \times SU(q) = G/Z_p, \quad SU(p) \times U(q) = G/Z_q,$$
$$S(U(p) \times U(q)) = G/Z_{p+q}.$$

The identifications of elements which define the quotient groups, are, respectively,

$$(\omega = \omega'), \quad (\sigma = \sigma'), \quad (\omega = \omega', \sigma = \sigma'), \qquad (5.24)$$

where $\omega \in Z(SU(p))$, $\sigma \in Z(SU(q))$, ω', σ' are elements of $U(1)$. In the CUIR $(\mu(p), \mu(q), m)$ these elements are represented by $\omega^{t(p)}$, $\sigma^{t(q)}$ and $(\omega')^m$, $(\sigma')^m$ respectively, and hence the quotient groups are the true groups of the representation if

$$m = t(p) \,(\bmod\, p), \qquad m = t(q) \,(\bmod\, q),$$

$$m = t(p) \,(\bmod\, p) \quad and \quad m = t(q) \,(\bmod\, q), \tag{5.25}$$

respectively. In the defining, $(p+q)$-dimensional representation of $S(U(p) \times U(q))$ these conditions are satisfied if $U(1)$ is represented as $e^{i\phi}1_p + 1_q$, $1_p + 1_q e^{i\phi}$ and $e^{i\phi}(1_p + 1_q)$ respectively.

For physics, the relevant examples of $SU(p) \times U(1)$ at the present time are $SU(2) \times U(1)$, which occurs both as a strong (flavour) symmetry algebra and as the electroweak gauge algebra, and $SU(3) \times U(1)$ which occurs as the maximal unbroken gauge-symmetry algebra. In all of these three occurrences the condition (5.23) is satisfied for all known physical states, including quarks (chapter 6) and thus it appears that the global groups which are realized in nature are $U(2)$ and $U(3)$. The relevant example of $SU(p) \times SU(q) \times U(1)$ is $SU(3) \times SU(2) \times U(1)$, which is the algebra of the strong and electroweak gauge interactions. In turns out that in this case the strongest condition $m = t(p)$, $m = t(q)$ in (5.25) is satisfied by all known physical states, including quarks (chapter 6). Thus it appears that the global gauge group of the strong and electroweak gauge interactions is $S(U(3) \times U(2))$. It is interesting to note that the global groups $U(2)$, $U(3)$ and $S(U(2) \times U(3))$ occur naturally when larger simple groups are spontaneously broken (chapter 11). Thus the occurrence of these global groups supports the hypothesis that larger groups than the observed ones exist. (Of course, in the case of flavour $U(2)$ this hypothesis has long since been verified by the existence of flavour $SU(3)$.)

5.4 Reality properties of the CUIRs

An important question for grand unification is whether the CUIRs of the simple compact Lie groups are real. There are actually three kinds of reality properties, namely strictly real, pseudo-real and strictly complex, and they are defined as follows: Let $U^*(g)$ be the complex conjugate of a representation $U(g)$. If $U^*(g)$ is not unitarily equivalent to $U(g)$ then $U(g)$ is said to be strictly complex. If $U^*(g)$ *is* unitarily equivalent to $U(g)$ then there exists a constant unitary matrix W such that $U^*(g) = W^\dagger U(g) W$, $W^\dagger W = 1$, and it can be shown (exercise 5.3) that $W^*W = \pm 1$. If $W^*W = 1$ a basis can be found in which $U(g)$ is a real matrix, $U^*(g) = U(g)$, and the representation is said to strictly real. If $W^*W = -1$, on the other hand, there is no basis in which $U(g)$ is a real matrix. In this case $U(g)$ is said to be pseudo-real. A well-known example of a pseudo-real representation is the defining, two-dimensional representation of $SU(2)$, $U(\mathbf{a}) = \exp i\mathbf{a}\cdot\boldsymbol{\sigma}$ where $\boldsymbol{\sigma}$ are the Pauli matrices. This representation cannot be made real since otherwise it would be a representation of $SO(2)$ which is abelian, but nevertheless $U(g) = CU^*(g)C^{-1}$ where $C_{ij} = \epsilon_{ij}$. Note that, for the vectors $\mathbf{v}$ in the representation space of a pseudo-real representation, the pseudo-reality property $\mathbf{v}^* = W^{-1}\mathbf{v}$ is compatible with the group transformations.

Since the CUIRs of the compact simple groups are the (unique) leading CUIRs in the products of the primitive representations, their reality properties are determined by the primitive CUIRs. Hence it suffices to consider the reality properties for the primitive representations and for the different Cartan classes. They are as follows:

$SU(n)$: $n \geqslant 3$: The fundamental representation F is strictly complex and the l primitive representations F_s satisfy the condition $F_s^* = F_{l-s}$. Thus they are all strictly complex except $F_{\frac{1}{2}l}$ for even l, which is strictly real. It follows that all the CUIRs are strictly complex, except those for which

$$(\mu_1\mu_2...\mu_{l-1}\mu_l) = (\mu_l\mu_{l-1}...\mu_2\mu_1)$$

which are strictly real.

$Sp(2n)$ (*including* $SU(2)$): The symplectic and unitary conditions $U^\dagger(g)U(g) = 1$, $\tilde{U}(g)JU(g) = J$ where $J^2 = -1$, $J = J^\dagger$ imply that $U^*(g) = J^\dagger U(g)J$ where $J^\dagger J = 1$, $J^*J = -1$ so the fundamental representation F is pseudo-real. Since the product of two pseudo-real representations is strictly real, the odd- and even-order fundamental representations F_{2s+1} and F_{2s} are pseudo-real and strictly real respectively. Similarly for the odd- and even-order tensors of any rank.

$SO(n)$: Since the defining representations are real, all the tensor representations are real (except for the irreducible parts $F_l^{\pm}$ of F_l for $SO(4n+2)$). In order to study the spinor representations one must consider first the reality properties of the Clifford algebras. Since the Pauli matrices are pseudo-imaginary, $\sigma_i^* = -\sigma_2\sigma_i\sigma_2^{-1}$, it follows from exercise 4.4 that the Clifford matrices γ_μ are pseudo-imaginary, $\gamma_\mu^* = -\sigma\gamma_\mu\sigma^{-1}$, where $\sigma = \sigma_2 \times \sigma_2 \times \sigma_2 \times \sigma_2...$, for $\mu = 1, ..., 2p$, and γ_0 is pseudo-imaginary or imaginary according as p is odd or even. One has therefore the following situation:

$SO(2l+1), SO(2l) = SO(4n)$: the spinor representations $\Delta, \Delta^{\pm}$ are pseudo-real, $\Delta^* = \sigma\Delta\sigma^{-1}$, $(\Delta^{\pm})^* = \sigma\Delta^{\pm}\sigma^{-1}$.

$SO(2l) = SO(4n+2)$: the spinor representations $\Delta^{\pm}$ are strictly complex, but are conjugate $(\Delta^{\pm})^* = \sigma\Delta^{\mp}\sigma^{-1}$.

For the exceptional groups all primitive representations are real except the fundamental 27-dimensional representation of E_6 (Tits, 1967), which is strictly complex.

For many grand unification theories the fermion representations are required to be strictly complex, and for such theories one sees that the groups $SU(n)$, $SO(4n+2)$ and E_6 are the only candidates. Furthermore, for $SO(4n+2)$, and E_6 the representations $\Delta^{\pm}$ and 27 (or their product with real representations) must be used.

5.5 Dimension of the CUIRs

From the weight diagrams it is evident that the dimension of a CUIR is simply the number of weights, with multiplicities $\gamma(m)$ included,

$$\dim \mathbf{j} = \sum_m \gamma(m). \tag{5.26}$$

However, this formula is not particularly useful, since to evaluate it one needs to compute all the multiplicities. A useful formula, which expresses the dimensions in terms of the highest weights j, may be obtained by taking the limit $\phi \to 0$ in Weyl's character formula. The result (see, for example, Humphreys, 1972) is

$$\dim(\mathbf{j}) = \prod_{\alpha > 0} \frac{(j+\delta, \alpha)}{(\delta, \alpha)}, \tag{5.27}$$

where the product is over all positive roots and 2δ is the sum of the positive roots. Thus for $SU(2)$ and $SU(3)$ one has

$$\dim(\mathbf{j}) = 2j+1, \quad \dim(\mathbf{j}) = (j_1+1)[(j_1+1)^2 - j_2^2]$$
$$= (I+1)[(I+1)^2 - \tfrac{9}{4}Y^2], \tag{5.28}$$

respectively, where I is the conventional isospin and Y the conventional hypercharge. An even more useful formula, which expresses the dimension in terms of the Dynkin indices μ, may be obtained by expanding α in terms of the primitive roots, and δ and j in terms of the primitive weights,

$$\alpha = \sum_{s=1}^{l} n_s \alpha_s, \quad \delta = \sum_{s=1}^{l} \omega_s, \quad \mathbf{j} = \sum_{s=1}^{l} \mu_s \omega_s, \tag{5.29}$$

(see table 2 for δ) and using this result one obtains from (5.27) the formula

$$\overline{\dim}\,(\mu) = \prod_\alpha \frac{\sum\limits_{s=1}^{l} n_s(\mu_s+1)}{\sum\limits_{s=1}^{l} n_s}, \qquad (5.30)$$

where the n_s depend only on the group and the Dynkin indices μ_s only on the representation. For example, for the rank-2 groups $SU(3)$, $SO(5)$, G_2, the formula (5.30) yields

$$\dim\,(\mu) = \frac{ab(a+b)}{2},\quad \frac{ab(a+b)(2a+b)}{3!},\quad \frac{ab(a+b)(a+2b)(2a+3b)}{5!},$$

$$(5.31\,a)$$

respectively, where $a = \mu_1+1$, $b = \mu_2+1$.

The formula (5.30), which is valid for all simple compact Lie groups, can be further simplified by specializing to the different Cartan classes. For example for $SU(n)$ it reduces to

$$\dim\,(\mu) = \prod_{s=0}^{l-1} \frac{A(s)}{(s+1)!} \qquad (5.31\,b)$$

where

$$A(s) = \prod_{k=1}^{l-s} (a_k+a_{k+1}+\ldots+a_{k+s}) \quad \text{and} \quad a_k = \mu_k+1.$$

The reduction for each of the classical groups has been discussed in detail by Antoine and Speiser (1964).

In most cases the dimension is already enough to characterize the CUIR because (5.30) has a unique solution for non-negative integers μ_s. But this is not always the case. For example, the $(4, 0)$ and $(2, 1)$ representations of $SU(3)$ are both 15-dimensional and, as has been seen, $SO(8)$ has three inequivalent 8-dimensional CUIRs.

5.6   Casimir invariants and indices

A concept that is useful for Lie algebras, especially for linear representation theory, is that of the enveloping algebra, which is simply defined as the linear span of the (associative) symmetric products $X_a X_b$, $X_a X_b X_c$, ... (The antisymmetric products are not linearly independent because of the commutation relations.) For linear representations the above products become simply the ordinary symmetric products $U(X_a)\,U(X_b)$, $U(X_a)\,U(X_b)\,U(X_c)$ ... of the representative matrices. An important property of the enveloping algebra is that it admits non-trivial central elements, i.e. elements that commute with all the elements of the Lie algebra). According to Schurs lemma the central elements are multiples of the identity in any irreducible representation.

The most useful of the central elements is the second-degree Casimir element $C_2 = -g^{ab}X_a X_b$, where g^{ab} is the Cartan metric. For representations it is convenient to normalize C_2 as

$$C_2(j) = g^{ab}\sigma_a \sigma_b, \quad g^{ab}g_{bc} = \delta_c^a, \quad g_{bc} = \operatorname{tr} F_a F_b, \quad (5.32)$$

where the σs are the generators of an arbitrary representation **j** and the Fs the generators of the fundamental representation because then its value is independent of the normalization of the Fs. In particular $C_2(F) = \dim G/\dim F$. The operator C_2 was first introduced by Casimir in order to study the reducibility of the representations of $SO(3)$, but by using the Jacobi relation it is easy to verify that it is central for any Lie algebra. For semi-simple Lie algebras, C_2 can be written as a sum of squares with coefficients $g_{ab} = \pm\delta_{ab}$, and for compact semi-simple Lie algebras, with coefficients $g_{ab} = -\delta_{ab}$ (exercise 5.4).

Since C_2 is a multiple of the identity in any hermitian irreducible representation (HIR) of the algebra its value may be obtained by letting it operate on the highest weight, or, more simply by taking its expectation value with respect to the highest weight. From the Cartan basis for the generators one easily obtains

$$C_2/2 = (j \mid H_i^2 + \{E_\alpha, E_{-\alpha}\} \mid j) = j_i^2 + \alpha_i j_i = (j, j + 2\delta) = (j + \delta)^2 - \delta^2, \quad (5.33)$$

where, as usual, 2δ is the sum of the positive roots and where the δ are normalized so that $C_2(F) = \dim G/\dim F$. For $SU(2)$ the expression (5.33) reduces to the well-known expression $j(j+1)$.

The Casimir operator plays an important role in the renormalization group properties of non-abelian gauge theories (see section 7.5 on asymptotic freedom). The quantity that actually enters in that case is not $C_2(j)$ itself but the related quantity $I_2(j)$ already defined in (5.17) as

$$\operatorname{tr} \sigma_a \sigma_b = I_2(j) g_{ab}, \quad \text{or} \quad I_2(j) = \frac{\dim(j) C_2(j)}{\dim(F) C_2(F)} = \frac{\dim(j)}{\dim(G)} C_2(j). \quad (5.34)$$

The quantity $I_2(j)$ is called the *index* of the representation and is an integer for all representations, including spinor representations except for $SO(n)$, $n \leqslant 6$ (see table 3). Note $C_2(F) = \dim G/\dim F \Leftrightarrow I_2(F) = 1$.

From (5.33) one sees that the value of the second-order Casimir is quantized and that for $SU(2)$ its value determines the representation (since there is only one positive solution of $j(j+1) = \frac{1}{4}n(n+2)$, n integer). For

Table 3. *The second-degree Casimirs* $C_2(j) = 2(j, j+2\delta)$ *and indices* $I_2(j) = C_2(j)\,\dim(j)/\dim G$ *are given for the primitive and adjoint representations of the classical groups. The Casimir is normalized so that* $I_2(F) = 1$ *for the fundamental tensor representation F, the symbol δ denotes half the sum of the positive roots (which, together with j, is computed from table 2)* $e = e_1 + e_2 + \ldots + e_n$, $E = ne_1 + (n-1)e_2 + \ldots + 2e_{n-1} + e_n$ *and* $n = l$ *except for $SU(n)$ where $n = l+1$. Note that δ is also the sum of the primitive weights.*

	$SU(n)$	$Sp(2l)$	$SO(2l+1)$	$SO(2l)$
δ	$E - \dfrac{n+1}{2}e$	E	$E - \tfrac{1}{2}e$	$E - e$
$\dim(F_k)$	$\dbinom{n}{k}$	$\dbinom{2l}{k} - \dbinom{2l}{k-2}$	$\dbinom{2l+1}{k}$	$\dbinom{2l}{k}$
$C_2(F_k)$	$\dfrac{(n+1)}{n}k(n-k)$	$\tfrac{1}{2}k(2l+2-k)$	$\tfrac{1}{2}k(2l+1-k)$	$\tfrac{1}{2}k(2l-k)$
$I_2(F_k)$	$\dbinom{n-2}{k-1}$	$2\dbinom{2l-1}{k-1}\left(\dfrac{l-k+1}{2l-k+1}\right)$	$\dbinom{2l-1}{k-1}$	$\dbinom{2l-2}{k-1}$
$C_2(\text{adj})$	$2n$	$2(l+1)$	$(2l-1)$	$2(l-1)$
	$C_2(\Delta) = \tfrac{1}{8}(2l+1)$	$I_2(\Delta) = 2^{l-3}$	$C_2(\Delta^{\pm}) = \tfrac{1}{8}(2l-1)$ $I_2(\Delta^{\pm}) = 2^{l-4}$	

higher rank groups this cannot be true since j is replaced by an l-vector $\mathbf{j}$, but in that case one might expect to have l Casimirs to determine the l components of $\mathbf{j}$. This turns out to be the case (Racah, 1965) and the l Casimirs turn out to be the l invariants

$$C_k(j) = g^{ab\ldots rs}\sigma_a\sigma_b\ldots\sigma_r\sigma_s, \quad g^{ab\ldots rs} = \mathrm{tr}\,(F^aF^b\ldots F^rF^s)_{TS}, \quad (5.35)$$

of lowest non-vanishing degree, where $F^a = g^{ab}F_b$ denotes the generators of the fundamental representation of the group in question, and TS denotes total symmetrization. The lowest-degree non-vanishing independent C_k for the various groups are those with $k = 2, \ldots, l+1$ for $SU(n)$, $k = 2, 4, 6, \ldots, 2l$ for the other classical groups and $k = 2, 6$ for G_2. For $SO(2l)$ the invariants are not quite complete in that they do not distinguish between Δ^+ and Δ^- like representations, but they can be made complete by eliminating C_{2l} in favour of the Levi-Civita invariant

$$E_l = \epsilon_{ijrs\ldots ab}\sigma_{ij}\sigma_{rs}\ldots\sigma_{ab}, \quad (E_l^2 \approx C_{2l}), \quad (5.36)$$

where σ_{ij} denotes the generators in antisymmetric matrix notation.

The C_k have the property that if they are expressed in terms of $\mathbf{j}+\delta$ by writing $C_k(j+\delta) = \langle j|C_k|j\rangle$ as in (5.33) then they are invariant with respect to the Weyl group (just as $(j+\delta)^2$ in (5.33) is reflexion invariant). An important use of this property is to prove the completeness of the C_k, i.e. to show that they determine the highest weights **j** uniquely. The proof (Racah, 1950) is based on the fact that for each group the product p of the degrees of the C_k (e.g. $p = (l+1)!$ for $SU(n)$) is just the order of the Weyl group, and hence only one of the p permitted solutions of $C_k = $ constant, $k = 1, ..., l$ can be a dominant (and hence a highest) weight.

For gauge theories the most important Casimir next to C_2 is the third-degree one, C_3, which occurs as a coefficient in the axial anomaly (section 7.4). Actually, as in the case of C_2, the quantity that enters directly is not C_3 itself but the associated *index* I_3, defined, in analogy to (5.34) as

$$\tfrac{1}{2}\mathrm{tr}\, \sigma_a\{\sigma_b, \sigma_c\} = I_3(j)\, g_{abc}, \tag{5.37}$$

where g^{abc} are the coefficients of $C_3(j)$ in (5.35). It is easy to verify that, in analogy to (5.34),

$$I_3(j) = \frac{\dim(j)\,C_3(j)}{\dim(F)\,C_3(F)} = \frac{\dim(j)}{(g \cdot g)}\,C_3(j), \tag{5.38}$$

where $(g \cdot g)$ denotes $g^{abc}g_{abc}$ and is independent of the normalization of the Fs.

Exercises

5.1. Show that the transformations $z \to z' = az + b/cz + d$, where $(ad - bc) = 1$ form a nonlinear representation of $SL(2, C)$ on the complex plane.

5.2. Find the true groups for the HIRs (μ, m) of $SU(6) \times U(1)$ and $SU(4) \times SU(2) \times U(1)$ (i.e. p, q not prime).

5.3. Show that if the CUIR $U(g)$ is equivalent to its complex conjugate, $U^*(g) = W^\dagger U(g)W$, $W^\dagger W = 1$, then $W^*W = \pm 1$, and that if $W^*W = 1$ there exists a basis in which $U(g)$ is real. (Hint: use $\sqrt{W}$.)

5.4. Use the Cartan canonical form to show that for compact simple algebras the second-order Casimir may be written as a positive-definite sum of squares.

5.5. Show that when the roots are normalized as in table 2 (p. 41) the highest weight in the symmetric tensor representation of $SU(n)$ of rank k is $k(e_1 - e_0)$. Hence, or otherwise, show that the second-order Casimir and index of this representation are respectively

$$C_2 = \left(\frac{n-1}{n}\right)(n+k)k \quad \text{and} \quad I_2 = \binom{n+k}{n-1}.$$

6
Rigid internal groups

6.1 Introduction

Part II of the monograph will be concerned with gauge theory and spontaneous symmetry breaking, and since both of these subjects are based on rigid internal group theory this last chapter of Part I will be devoted to that subject in order to prepare the ground. In their broadest sense, internal rigid groups are space–time independent compact Lie groups which are used to classify physical fields, or particles, and various physical quantities, such as charges, currents, masses and magnetic moments, associated with them. The term 'classification' here means the assignment of some finite multiplet of such quantities, the fields $\Phi_a(x)$ themselves, say, to a unitary representation $U(g)$ of the group, so that the group transformation reads

$$\Phi(x) \to U(g)\,\Phi(x) \quad \text{or} \quad \Phi_a(x) \to U_{ab}(g)\,\Phi_b(x), \tag{6.1}$$

where $(\Phi_a, \Phi_b) = \delta_{ab}$, and $a = 1, ..., n$. The fields are supposed ortho-normal, as shown, and it is in order to preserve this orthonormality that the representation is required to be unitary and the group compact. Physically, the preservation of orthonormality corresponds to the group invariance of the canonical commutation relations and the kinetic part of the Lagrangian, both of which are described by positive-definite hermitian forms.

It is *not* assumed that the whole Lagrangian L, or equivalently, its interaction part, is invariant with respect to (6.1), i.e. that G is a symmetry group of the interactions. However in many cases G is at least an approximate symmetry group of L, and in these cases it is often referred to as an approximate internal symmetry group. In a few cases, notably the case of the isospin group in traditional nuclear physics without electromagnetic interactions, it is even an exact symmetry group, $L(U(g)\,\Phi) = L(\Phi)$.

With respect to space–time, the fields $\Phi(x)$ are assumed, of course, to transform covariantly with respect to the Poincaré group,

$$(U(a, \Lambda)\,\Phi)\,(x) = D(\Lambda)\,\Phi\,(\Lambda^{-1}(x-a)), \tag{6.2}$$

where (a, Λ) are the parameters of translations and Lorentz transformations respectively, and $D(\Lambda)$ is the relevant representation of the Lorentz group. In practice only renormalizable theories are considered, so the $\Phi(x)$ are scalars, for which $D(\Lambda) = 1$, or spin-$\frac{1}{2}$ fermions, for which $D(\Lambda)$ is the 4×4 Dirac, or 2×2 Weyl representation of the Lorentz group. The most general renormalizable Lagrangian that can be written for such fields is

$$-L = \bar{\psi}\not{\partial}\,\psi + \tfrac{1}{2}(\partial_\mu \phi)^2 + V(\phi) + \{m_{\alpha\beta}\bar{\psi}_\alpha \psi_\beta + g^a_{\alpha\beta}\bar{\psi}_\alpha \psi_\beta \phi_a$$
$$+ f^k_{\alpha\beta}\bar{\psi}_\alpha \gamma_5 \psi_\beta \phi_k + \text{herm. conj.}\}, \tag{6.3}$$

where the ϕ_a, ϕ_k and ψ denote any number of scalar, pseudo-scalar and spin-$\frac{1}{2}$ fields respectively (Weyl fields χ are included as Dirac fields ψ with the subsidiary chiral condition $(1 \pm \gamma_5)\,\psi = 0$), m, g, f are constants (Dirac masses and couplings) and $V(\phi)$ is a polynomial of fourth degree in ϕ, which is bounded below. The gauge principle will be based on such Lagrangians. When the Lagrangian is invariant with respect to (6.1) the constants $m_{\alpha\beta}$, $g^a_{\alpha\beta}$, $f^a_{\alpha\beta}$ and the constants in $V(\phi)$ become invariant tensors (Levi-Civita symbols, CG-coefficients etc.). For example, for a scalar field ϕ and a Dirac field ψ assigned to the three- and two-dimensional representations of $SO(3)$ respectively, the most general $SO(3)$-invariant renormalizable Lagrangian would be

$$-L = \tfrac{1}{2}(\partial_\mu \phi)^2 + \bar{\psi}_\alpha(\not{\partial} - m)\,\psi_\alpha + g\bar{\psi}_\alpha \sigma_{\alpha\beta}\psi_\beta \cdot \phi + \lambda(\phi \cdot \phi)^2 + \mu(\phi \cdot \phi), \quad \lambda > 0, \tag{6.4}$$

where σ are the Pauli matrices. When the Lagrangian is only approximately invariant, the constants deviate from invariants by small amounts, e.g.

$$g\bar{\psi}\sigma\psi \cdot \phi \to g\bar{\psi}\sigma\psi \cdot \phi + h\bar{\psi}\sigma_3 \psi \phi_3, \tag{6.5}$$

where $h \ll g$. When only the mass terms are not symmetric (whether by a small amount or not) the symmetry is said to be softly broken.

In general, the representation (6.1) is not the same for the different fields ψ, χ, ϕ, and is not irreducible for any of these sets of fields alone. However, it is clear that any symmetry will reduce the number of independent constants in the Lagrangian, and the more irreducible the representation the greater the reduction will be.

Because the internal symmetry is rigid, i.e. g does not depend on x, one has

$$\partial_\mu U(g)\,\Phi(x) = U(g)\partial_\mu \Phi(x), \tag{6.6}$$

where $\Phi = \phi$, ψ or χ, and thus the fact that the Lagrangian contains derivatives makes no difference to its symmetry properties. In fact, as mentioned above, the kinetic part, which is the only part that contains derivatives, is always symmetric,

$$L_{\text{kin}}(\Phi, \partial_\mu \Phi) = L_{\text{kin}}(U(g)\, \Phi, \partial_\mu U(g)\, \Phi). \tag{6.7}$$

6.2   Noether currents

One of the most important consequences of continuous rigid symmetries is the existence of conserved charges. These charges are called Noether charges and are constructed as follows: Let $\Phi(x)$ denote $\phi(x)$ or $\psi(x)$ and I_k the infinitesimal generator of the representation $U(g)$,

$$\left(\frac{\partial}{\partial a_k} (U(g)\, \Phi(x))^\alpha \right)_{a=0} = (I_k\, \Phi(x))^\alpha = (I_k^{\alpha\beta})\, \Phi^\beta(x), \tag{6.8}$$

where I_k is just a numerical matrix. Then let j_μ^k be the current

$$j_\mu^k = \pi_\mu I_k \Phi = \pi_\mu^\alpha I_k^{\alpha\beta} \Phi^\beta \tag{6.9}$$

where

$$\pi_\mu = \frac{\partial L(x)}{\partial \Phi_\mu(x)}, \quad \Phi_\mu = \partial_\mu \Phi.$$

It is easy to verify that, on account of the Euler–Lagrange equations of motion, one has

$$\partial_\mu j_\mu^k = (\partial_\mu \pi_\mu) I_k \Phi + \pi_\mu I_k (\partial_\mu \Phi) = \frac{\partial L}{\partial \Phi} I_k \Phi + \frac{\partial L}{\partial \Phi_\mu} I_k \Phi_\mu = \left(\frac{\partial L}{\partial a_k} \right)_{a=0}. \tag{6.10}$$

Hence if L is symmetric, this current is conserved

$$\frac{\partial L}{\partial a_k} = 0 \Rightarrow \partial_\mu j_\mu^k = 0. \tag{6.11}$$

Thus for each generator of the continuous internal symmetry group there is a conserved current. If one now defines the (Lorentz-invariant) charges

$$Q_k = \int \mathrm{d}^3 x j_0^k(x), \tag{6.12}$$

one sees from (6.11) that (with suitable boundary conditions at spatial infinity) these charges are conserved,

$$\frac{\partial}{\partial t} Q_k = 0. \tag{6.13}$$

The Q_k are the required Noether charges. Note that they are conserved only by virtue of the equations of motion.

The Noether charges have the further property that if the system is quantized in the canonical manner

$$[\pi_0(x), \Phi(y)] = i\delta^3(x-y), \quad (x_0 = y_0), \tag{6.14}$$

where [,] denotes commutator or anticommutator (or indeed Poisson bracket), and all other equal-time fields commute (or anticommute), then the Q_k are generators of the symmetry in the sense that

$$[Q_k, \Phi(x)] = \int d^3y[\pi_0(y), \Phi(x)] I_k \Phi(y) = iI_k \Phi(x). \tag{6.15}$$

Equation (6.15) implies also, of course, that

$$[Q_p, Q_q] = if_{pq}^s Q_s, \tag{6.16}$$

where f_{pq}^s are the structure constants. Thus the Noether charges are the generators of the (infinite-dimensional, reducible) representation of G carried by the fields. Since the Noether charges are therefore elements of a Lie algebra, they are additive, i.e. the charges for the product of two representations are the sums of the charges for the two representations $[\exp iQ_k \exp iQ'_k = \exp i(Q_k + Q'_k)]$. This means that the eigenvalues of the charges are *additive* quantum numbers, i.e. the numbers for a composite system are the sums of the numbers for the components. The additivity holds no matter what the binding force is, and for this reason is of great physical importance. For example, the fact the quarks have fractional electrical charge (section 6.3) is a consequence of this fact.

6.3 Application of the rigid internal groups

In order to illustrate the use of rigid internal groups and to prepare the ground for the introduction of gauge theory, the use of these groups for the observed elementary particles, or fields, will next be considered. By 'elementary' fields are meant here the fields associated with all the traditional elementary particles, the hadrons (mesons and baryons) and the leptons, where the hadrons are all the strongly interacting particles (the mesons and baryons being the bosonic and fermionic hadrons respectively) and the leptons the fields that have no strong interactions. The complete list of observed fields is given in the (annual) *Review of Particle Properties* (Wohl, 1984). There are actually two ways of regarding the rigid internal groups for elementary particles, corresponding to two stages in the historical development. From the first point of view (earlier historical period) the rigid groups are simply regarded as empirical classification groups for the observed strongly interacting fields and their properties. From this classification the existence of *quarks* is inferred, where quarks are subparticles out of which the observed particles are composed. From the second (present) point of view the quarks are regarded as the starting point, and the rigid internal groups and their properties are considered as consequences of the quark model. In the present section the first stage is recalled, and in the following section the quark model is briefly reviewed.

The first rigid symmetry of the hadrons to be observed is the fact that the strong nuclear forces of the proton (p) and neutron (n) are independent of electric charge (i.e. p–n, p–p, n–n forces are observed to be the same, modulo Fermi statistics). This symmetry is expressed by introducing a rigid symmetry group $SU(2)$, called isospin, assigning the nucleons $N = $ (p, n) to the two-dimensional representation of this $SU(2)$, and assuming the strong-interaction Lagrangian is $SU(2)$-invariant, i.e. is formed from inner products of the form $\bar{N}N$, $(\bar{N}\sigma N)\cdot(\bar{N}\sigma N)$, etc. Isospin is an exact symmetry group of nuclear physics when weak and electromagnetic interactions are neglected. This remains true for excited nuclear states, provided the isospin assignment is extended in an appropriate manner. For example, for the metastable hadrons (Wohl, 1984), one has $\{\pi^0, K^0\} = \{$triplet, doublet$\}$ for the pseudo-scalar mesons and $(\Sigma^{\pm 0}, \Lambda, \Xi^0) = \{$triplet, singlet, doublet$\}$ for the spin-$\frac{1}{2}$ baryons. Actually when the full roster of metastable hadrons is included, the exact symmetry group of the strong interactions turns out to be slightly larger than $SU(2)$, namely $U(2)$. The additional Noether charge for the central $U(1)$ in $U(2)$ is called hypercharge Y, and its assignments for $\{\pi, K\}$ are $\{0, 1\}$ and for $\{N, \Sigma, \Lambda, \Xi\}$ are $(1, 0, 0, -1)$. For all hadrons and resonances, hypercharge is given by the equation

$$Q = T_3 + \tfrac{1}{2}Y, \qquad (6.17)$$

where T_3 is the third (diagonal) component of isospin and Q is the electric charge. The hypercharge defined by (6.17) turns out to be an integer for all observed fields. Because of this and the fact that the electric charge is an integer for all observed fields, $Y + 2T_3$ is even, and this is just the condition of section 5.5 for the group to be $U(2) = SU(2) \times U(1)/Z_2$ rather than $SU(2) \times U(1)$. In addition to $U(2)$ the strong interactions have an exact discrete symmetry group Z_2 which connects particles and antiparticles (Michel, 1953). The non-trivial element of this group is $C \exp(i\pi T_2)$ where C is the charge-conjugation operator and T_2 is the second generator of isospin. This element is called the isoparity, G, and the observed fields are eigenstates of G with eigenvalues ± 1 (Lee and Yang, 1956).

An intriguing feature of the hadron multiplets $\{\pi, K\}$, $\{N, \Sigma, \Lambda, \Xi\}$ and higher excitations is that, in spite of the fact that they are reducible with respect to $U(2)$, the spins and parities are uniform within the multiplets (0^- for $\{\pi, K\}$, $\frac{1}{2}^+$ for $\{N, \Sigma, \Lambda, \Xi\}$ etc.) and the masses of the baryons deviate from the average by less than ten per cent. Furthermore, the electroweak properties of the fields within a given multiplet, for example the weak decay rates and the magnetic moments of $\{N, \Sigma, \Lambda, \Xi\}$, exhibit similar regularities. It is found (Gell-Mann and Ne'eman, 1964) that all these regularities may be accounted for by embedding $U(2)$ in an $SU(3)$ group, called flavour $SU(3)$, or $SU(3, f)$, and assigning, not only the hadrons themselves, but their mass matrices, and weak and electric currents to appropriate (singlet, octet and decuplet) representations of this group. In fact further regularities of the fields are predicted in this manner. For example, the assignment of the mass matrix to a mixture of a singlet and octet representation leads to the well-known mass formula

$$M = a + bI + c[I(I+1) - \tfrac{1}{4}Y^2], \qquad (6.18)$$

for hadrons in any irreducible representation of $SU(3, f)$, where a, b, c are constants which depend only on the representation, and the assignment of the electromagnetic current to an octet allows the magnetic moments of any baryon multiplet to be expressed in terms of one or two parameters. Thus the magnetic moments of Σ, Ξ and the transition rate $\Sigma_0 \to \Lambda$ can be expressed as linear combinations of the magnetic moments of the nucleons. Perhaps the most important prediction of $SU(3, f)$, however, is the existence of new fields e.g. a $U(2)$-singlet pseudo-scalar η to partner $(\pi, K, \bar{K})$ in an $SU(3)$-octet, and a $Y = 2$, $SU(2)$-singlet spin-$\frac{3}{2}$ particle Ω^- to partner the resonances (N^*, Σ^*, Ξ^*) in an $SU(3, f)$ decuplet. The experimental discovery of Ω^- within a few per cent of the mass predicted by (6.18) is generally regarded as one of the most spectacular successes of $SU(3, f)$.

The mass formula (6.18) implies, of course, that the masses within $SU(3, f)$ multiplets are not equal, and hence that $SU(3, f)$ cannot be an exact symmetry group for the hadrons, like $U(2)$. However, it is a good approximate symmetry, and an excellent classification group (Carruthers, 1971; Lichtenberg, 1970; Samios, Goldberg and Meadows, 1974). The most important of the $SU(3, f)$ classifications are:

metastable baryons: $J^p = \frac{1}{2}^+$ octet $\{N, \Sigma, \Lambda, \Xi\}$,

$\qquad\qquad\qquad J^p = \frac{3}{2}^+$ decuplet $\{N^*, \Sigma^*, \Xi^*, \Omega^-\}$

baryon resonances: $\frac{5}{2}^+$ octets, $\frac{3}{2}^+$ nonet mesons nonets: $0^- = \{(\eta)\,\eta_8,\,\pi,\,K,\,\bar{K}\}$, $1^+ = \{(\phi)\,\omega,\,\rho,\,K^*,\,\bar{K}^*\}$,

$$2^+ = ((f),f_8,\,A_2,\,K^{**},\,\bar{K}^{**}),$$

mass matrix: $M_0 + M_\eta^8$

electromagnetic current: $J_\mu^{em} = J_{\pi_3}^8 + \tfrac{1}{2}J_\eta^8$

charged weak current: $J_\mu^+ = \cos\theta_C\,J_{\pi^+}^8 + \sin\theta_C\,J_{K^+}^8$,

where the subscript denotes the position in the octet in terms of the η, π, K quantum numbers and θ_C is an empirical angle (Cabibbo, 1963).

There are three rather surprising features in these $SU(3,f)$ assignments:

(i) The absence of representations for which the triality t (section 5.3) is not zero, especially the absence of the defining representations 3 and 3* for the hadrons.

(ii) The occurrence of the mesons in nonets (octets+singlets), in which the physical fields (fields of definite mass) are linear combinations of the $SU(3,f)$-singlet and $U(2)$-singlet number of the octet.

(iii) The absence of any representations other than 0, 8, 10 for the fields and any representation other than 8 for the currents.

It is these three features of $SU(3,f)$ which were originally responsible for the introduction of the quark model, because in that model they have a simple and natural explanation. The model supposes that there exist quarks q, which are spin-$\tfrac{1}{2}$ fields in the fundamental (three-dimensional) representation of $SU(3,f)$, and that the observed mesons and baryons are bound states of $\bar{q}q$ and qqq respectively. This is because the spin-parity and $SU(3)$ decompositions of these products are

$$\left.\begin{aligned}
\bar{q}q &= 0^- \text{ nonet} + 1^+ \text{ nonet} &\text{(36 states),}\\
(qqq)_{TS} &= \tfrac{1}{2}^+ \text{ octet} + \tfrac{3}{2}^+ \text{ decuplet} &\text{(56 states),}
\end{aligned}\right\} \tag{6.19}$$

where TS denotes totally symmetric, and these states have exactly the quantum numbers of the two lowest-mass meson and baryon states respectively. Thus the quark model not only reproduces the $SU(3,f)$ assignments naturally but produces correlations between the $\tfrac{1}{2}^+$ and $\tfrac{3}{2}^+$ fields and the 0^- and 1^+ fields, respectively, which are absent in $SU(3,f)$. Similarly the octet currents can be formed from $\bar{q}\gamma_\mu q$ and $\bar{q}\gamma_\mu\gamma_5 q$, and the absence of higher-dimensional $SU(3,f)$ representations such as the 27-dimensional one can be explained by assuming that the resonances are similar states with higher *orbital*, but not internal, excitations.

The fact that the quark model reproduces the $SU(3,f)$ assignments in a natural manner is not, of course, sufficient in itself to establish its validity. But it forms a good basis for further investigation, and the further investigation that has been carried out, notably in baryon spectroscopy, current algebra and deep-inelastic scattering, put the quark model on a very firm footing (Close, 1979). Once the quark model is accepted, the first (inductive) phase in the history of the rigid symmetry groups of the strong interactions comes to an end. In the second (deductive) phase, which will be considered in the next section, the quark model is regarded as the primary construct, and the rigid internal group structures as consequences of it.

6.4 The quark model of the elementary fields

In the quark model the quarks are assumed to be spin-$\frac{1}{2}$ fermions and, in the absence of evidence to the contrary, are assumed to be local fields with point-like interactions. The two 'lightest' quarks $(u, d) = $ (up, down) are assumed to form a doublet with respect to isospin and the next lightest, the strange quark s, is assumed to be an isospin singlet. The three together form an $SU(3, f)$ triplet (u, d, s). Since the quantum numbers $2T_3$, Y and Q are *additive*, the requirement that they have the usual integer values for the baryons (qqq) means that their values for the quarks must be

$$T_3 = (\tfrac{1}{2}, -\tfrac{1}{2}, 0), \quad Y = (\tfrac{1}{3}, \tfrac{1}{3}, -\tfrac{2}{3}), \quad Q = (\tfrac{2}{3}, -\tfrac{1}{3}, -\tfrac{1}{3}), \qquad (6.20)$$

respectively, and thus Y and Q must be fractional relative to the baryonic values. The fact that the electric charge is fractional means that if they were free the quarks would be seen rather easily. The fact that they have not been seen has thus led to the assumption that they are *confined*, i.e. that the strong interactions have a mechanism which keeps them permanently bound. As will be discussed later, this mechanism is now thought to originate in the non-abelian gauge character of the strong interactions. For fields which are not free, the concept of mass is not well defined, but there are a number of empirical definitions of mass that can be used depending on the context, e.g. current masses in current algebra, constituent masses in hadron spectroscopy, or parameter masses (i.e. the values of the mass parameters) in the Lagrangian. The mass parameters for the u, d and s quarks are roughly 5, 8 and 150 MeV respectively (electron mass ≈ 0.5 MeV). From these masses it is evident that, at the quark level, $SU(3, f)$ is badly broken and, what is more surprising, even isospin is broken. It is then somewhat of a mystery why isospin is exact, and $SU(3, f)$ approximately exact, at the hadronic level. A possible explanation is that the bulk of the hadronic properties are due to the strong *interactions* (which are $SU(3, f)$ invariant) and only a small contribution comes from the

Lagrangian quark masses. Thus, for example, the tentative mass formula,

hadron mass = universal mass + sum of constituent quark masses,

is found to work quite well if the universal mass (which is thought to originate in the renormalization group) is of order of about 1 GeV. The smallness of the pion mass and the large π–K mass difference can also be explained in this way – namely as the vanishing of the coefficient of the universal mass due to the Goldstone mechanism for chiral symmetry breaking. These and related questions are discussed in detail by Gasser and Leutwyler (1982).

It turns out that when the electroweak interactions are taken into account the above, three-quark, picture needs to be extended. This is because the standard model of the electroweak interactions (section 9.3) requires that: (i) the quarks come in pairs (in order to couple to the weak gauge field W_μ^+); and (ii) each quark pair corresponds to a lepton pair (in order to make the theory renormalizable, as will be discussed in section 7.4). Accordingly, the existence of the (ν^e, e) (ν^μ, μ) lepton pairs requires the existence of two quark pairs and this, in turn, requires the existence of a fourth quark. The fourth quark is called the charmed quark c and (apart from a small rotation (section 9.4)) the electroweak pairs are assumed to be (u, d) and (c, s) respectively. The most immediate consequence of the existence of c is that there should exist corresponding new hadrons, namely mesons of the form $\bar{c}c$ (charmonium) and $\bar{c}q$, $\bar{q}c$, where $q = (u, d, s)$, and baryons of the form ccc, cqq, and ccq. Many of these hadrons have now been observed (Trilling, 1981) and have thus confirmed the existence of c (if not as a field, at least as a quantum number). In analogy to $SU(3, f)$ one may consider the four quarks (u, d, c, s) as belonging to the defining representation of a strong-interaction rigid internal group $SU(4, f)$ and construct baryons, currents, mass matrices, etc. as $SU(4, f)$ tensors. But even at the hadronic level $SU(4, f)$ is badly broken, so, in contrast to $SU(3, f)$, it is not even an approximate rigid symmetry group. The Lagrangian mass of the charmed quark is about ten times that of the strange quark, namely, $M_c \approx 1.2$ GeV. Thus c has approximately the same mass as the nucleons.

The more recent discovery of the τ-lepton, with (presumably) a neutrino partner ν to form a new lepton pair (ν^τ, τ), has necessitated the introduction of yet another quark pair $(t, b) = (\text{top, bottom})$ and, of course, for (t, b) the story of the charmed quark is repeated, i.e. the existence of (t, b) requires the existence of new mesons such as $\bar{t}t$, $\bar{b}q$, $\bar{q}t$, and of new baryons such as (ttt), (ttb), (tbq), (bqq), where q now denotes (u, d, c, s). A sufficient number of these hadrons have been seen to confirm the existence of b with a mass of approximately 5 GeV, and although earlier reports of t at about 40 GeV have not been confirmed it is generally believed that t exists and has a mass of at least that order. As in the case of (u, d, s, c) one may construct for (u, d, s, c, t, b) a rigid internal $SU(6, f)$ group, but it is by no means a symmetry group.

The quark–lepton pairs $\{(v^e, e) \sim (u, d)\}$ $\{(v^\mu, \mu) \sim (c, s)\}$ and $\{(v^\tau, \tau) \sim (t, b)\}$ are called *generations*, and one might well ask whether the process continues to include more generations, even an infinite number. The answer is not known for certain, but it is somewhat comforting to know that asymptotic freedom puts a fairly low limit on the number of generations according to section 7.6 and there are arguments from cosmology (notably upper bounds on the number of light neutrino species) that make the limit even lower (Dolgov and Zeldovich, 1981; Ellis, 1981; Steigman, 1979). In fact the cosmological arguments suggest four as the maximum number of generations, with three as the most likely, and if these arguments are correct the quark lepton picture is more or less complete. But of course, the occurrence of three separate generations, each a xerox copy of the others, except for the mass, remains a mystery. In fact, it is nothing but a generalization of the old electron–muon mystery.

The discussion of the rigid quark symmetries or classifications does not end with flavour. It turns out that the quarks have a further symmetry, called colour symmetry, which is exact and appears to be of much greater importance than flavour symmetry because it generates the strong inter-actions (section 9.2). The colour symmetry was originally introduced in order to avoid a problem of Fermi statistics raised by the baryon states $(qqq)_{\text{TS}}$. The problem is that these states are totally symmetric instead of totally antisymmetric (symmetric in flavour and spin because of the decomposition (6.19), and in configuration space because they are states of zero angular momentum). A natural way out of this problem is to assign a further index to the quarks, with respect to which the states qqq are totally antisymmetric, i.e. let $(u, d, s \ldots) \rightarrow (u^a, d^a, s^a \ldots)$ and $(qqq) = \epsilon_{abc} q^a q^b q^c$, $a = 1, 2, 3$. This new index is the colour, and it is clear that in order to form totally antisymmetric states there must be at least three colours. More direct evidence for the existence of colour symmetry is provided by processes such as the $\pi^0 \rightarrow 2\gamma$ decay rate and the ratio $R = \sigma(e^+ e^- \rightarrow \text{hadrons})/\sigma(e^+ e^- \rightarrow \mu^+ \mu^-)$, and suggests that there are just three colours, $a, b, c = 1, 2, 3$.

The colour symmetry is found to be exact and may be described by an exact rigid symmetry group $SU(m, c)$ where $m \geqslant 3$ and probably $m = 3$. This group acts only on the colour indices and is independent of flavour, a situation that may be expressed by saying that the full rigid internal group of the strong interactions is the direct product $SU(n, f) \times SU(m, c)$. Here $SU(n, f)$ is of course only a classification group (which becomes a more or less approximate symmetry group for hadrons when $n = 2, 3$) whereas $SU(m, c)$ is an exact rigid symmetry group. It should, perhaps, be emphasized that the existence of three colours ($m = 3$) has nothing to do with the 3 in $SU(3, f)$, and is due to the fact that baryons are of the form (qqq) no matter what the number of flavours. As already mentioned, the importance of colour symmetry is that (after gauging) it generates the strong interactions.

The final quark picture that emerges, therefore, is one in which the quarks have six flavours and three colours and thus belong to the defining representation rigid strong-interaction group $SU(6, f) \times SU(3, c)$, where $SU(6, f)$ is only a classification group, and $SU(3, c)$ is exact, i.e. the strong-interaction Lagrangian is exactly $SU(3, c)$-invariant. More generally the quarks may have $2k$ flavours and $m \geqslant 3$ colours, and belong to the defining representation of $SU(2k, f) \times SU(m, c)$. In the $SU(3, c)$ case the hadrons, or observable fields, are supposed to have quantum number corresponding to $\bar{q}^a q^a$ for mesons and $\epsilon_{abc} q^a q^b q^c$ for fermions. Thus, they are singlets with respect to $SU(3, c)$ but have all the quantum numbers that can be constructed from the products $\bar{q}q$ and qqq with respect to $SU(6, f)$ and spin (and orbital excitations). The only other fundamental fermions are the leptons, which are colour-singlets, and come in pairs that are in one–one correspondence with sextets of coloured quarks, e.g. $(\nu^e, e) \sim (u^a, d^a)$, $a = 1, 2, 3$. The octets consisting of a sextet of quarks and two leptons are called generations, and they are the natural groupings from the point of view of the electroweak interactions (section 9.3). The large number of fundamental fermions (at least eighteen quarks and six leptons) suggests that these fields are not really fundamental. But since attempts to replace them by more elementary fields (subquarks) have not been particularly successful, this is the picture that is used at present. Hence for the rest of this monograph, the quarks and leptons will be regarded as the fundamental fermions.

Exercises

6.1. Show that the classical Dirac Lagrangian
$$L = \bar{\psi}(\partial\!\!\!/ + m)\psi + (g\bar{\psi}\psi\phi + \text{herm. conj.}) + (\partial_\mu \phi)^2 + V(|\phi|^2)$$

is invariant with respect to the one-parameter group of chiral transformations

$$\psi \to e^{i\gamma_5 \alpha} \psi, \quad \phi \to e^{-2i\alpha} \phi$$

provided the fermion mass m is zero, and find the corresponding Noether current.

6.2. Okubo (1962) has shown that if the $SU(3)$ mass operator M is a mixture of an $SU(3)$-singlet and the $U(2)$-invariant member of an $SU(3)$ octet, it can be written as
$$M = \alpha + \beta\sigma_{33} + \gamma\sigma_{3i}\,\sigma_{i3}, \quad i = 1, 2, 3,$$

where α, β, γ are constants that depend only on the representation and σ_{ij} are the generators in matrix notation. Show that this formula is equivalent to (6.18).

6.3. Assuming that the quarks have spin $\frac{1}{2}$ and belong to the 3 of $SU(3)$, find the $SU(3)$ spin content of the three-quark states $(qqq)_{AS}$ that are totally *anti*symmetric in the combined $SU(3)$ spin indices, and verify that such states do not appear among the metastable particles in the tables of particles (Wohl, 1984).

6.4. Show that if the colour group were $SO(3)$ instead of $SU(3)$ there would exist unconfined (i.e. colourless) states with fractional electric charge. (Hint: consider the two-quark states qq.)

Part II
Gauge theory

7
The gauge principle

7.1 Introduction

As already mentioned, gauge theories are based on rigid internal groups, and the aim of the present chapter is to describe how (unbroken) gauge theories are constructed from rigid internal symmetry groups. The method of construction uses the so-called gauge principle (or minimal principle).

7.2 Electromagnetism

The simplest and most familiar case is that of electromagnetism, which is based on the rigid internal symmetry group $U(1)$, so for illustration this case is considered first. If the fields are as described in section 6.1 then the Lagrangian that describes their interaction with each other and with the electromagnetic (EM) field is simply

$$L_{EM} = L(\Phi, D_\mu \Phi) - \tfrac{1}{4} F_{\mu\nu} F^{\mu\nu}, \quad \Phi = (\phi, \psi, \chi), \qquad (7.1)$$

where $L(\Phi, \partial_\mu \Phi)$ is the usual Lagrangian that describes their interaction with each other,

$$F_{\mu\nu} = \partial_\mu A_\nu - \partial_\nu A_\mu, \quad (F_{i0} = E_i, \ F_{ij} = \tfrac{1}{2}\epsilon_{ijk} B_k, \ i,j = 1,2,3), \quad (7.2)$$

is the conventional EM field, and the *covariant* derivative D_μ is defined to be

$$D_\mu \Phi = \partial_\mu \Phi + iA_\mu Q\Phi, \qquad (7.3)$$

where $Q\Phi_a = e_a \Phi_a$ and e_a are the electric charges of the fields. Thus the EM Lagrangian differs from the matter Lagrangian (6.3) only in replacement of ∂_μ by D_μ and the insertion of the Maxwell kinetic term F^2. The principle $\partial_\mu \to D_\mu$ by which (7.1) is obtained from (6.3) is the gauge principle mentioned.

It is well known that the Lagrangian (7.1) is invariant with respect to gauge transformations of the first and second kind, namely the transformations

$$\Phi \to e^{i\theta Q}\Phi \quad \text{and} \quad \Phi \to e^{i\theta(x)Q}\Phi, \quad A_\mu \to A_\mu - \partial_\mu \theta(x), \quad 0 \leqslant \theta \leqslant 2\pi. \quad (7.4)$$

The invariance under gauge transformations of the first kind simply

77

expresses the fact that $L_{EM}(x)$ is electrically neutral, and is equivalent to the statement that $L_{EM}(x)$ is symmetric with respect to the rigid symmetry group $U(1)$ with parameter θ. The electric charges e_a are actually the group characters which define the (one-dimensional) representations of $U(1)$ to which the various fields belong, the electric current

$$j_\mu^{EM} = \frac{\delta L_{EM}}{\delta A_\mu} = e_\psi \bar\psi \gamma_\mu \psi + \mathrm{i} e_\phi \phi^* \overset{\leftrightarrow}{\partial}_\mu \phi + e_\phi^2 A_\mu \phi^* \phi, \qquad (7.5)$$

where the e_ϕ are the relevant charges, is the Noether current for this $U(1)$, and the electric charge operator is the corresponding Noether charge.

Gauge invariance of the second kind is quite different and marks the point of departure for gauge theories. It states that the Lagrangian is also invariant with respect to a symmetry group $U(1)$ which is not rigid, but varies (smoothly) from point to point in configuration space $\theta \to \theta(x)$. To obtain such invariance the introduction of the EM field is necessary since the kinetic terms for the matter fields do *not* remain invariant when $\theta \to \theta(x)$.

$$\partial_\mu \mathrm{e}^{\mathrm{i}\theta(x)Q} \Phi(x) \neq \mathrm{e}^{\mathrm{i}\theta(x)Q} \partial_\mu \Phi(x). \qquad (7.6)$$

In 1929 Weyl suggested making a virtue out of this necessity by reversing the usual point of view and using the requirement that symmetry be maintained when $\theta \to \theta(x)$ to motivate the introduction of electromagnetism. According to this point of view the potential A_μ is introduced *in order* to keep the derivatives of the matter fields covariant, i.e. so that

$$\mathrm{D}_\mu(A^\theta) \mathrm{e}^{\mathrm{i}\theta(x)Q} \Phi(x) = \mathrm{e}^{\mathrm{i}\theta(x)Q} \mathrm{D}_\mu(A) \Phi(x). \qquad (7.7)$$

The requirement (7.7) then determines the transformation property of the gauge potential A_μ, as given in (7.4). The kinetic term for A_μ is chosen to be invariant with respect to (7.4), and the other usual requirements, namely, that it be second order in the derivatives and Lorentz invariant, then determine it uniquely (apart from a constant factor). Invariance with respect to (7.7) also requires the mass term for A_μ to be zero, as indeed it is.

The process of letting $\theta \to \theta(x)$ is called 'gauging the $U(1)$ group' and the process of letting $L(x) \to L_{EM}(x)$ by introducing the gauge potential and the term F^2 is called 'gauging $L(x)$ with respect to $U(1)$'. Thus, according to Weyl, the EM field, and its interaction with matter, are consequences of gauging the matter Lagrangian with respect to $U(1)$. Of course, for EM theory, the gauge principle is no more than an interesting alternative to conventional EM theory (or to what used to be the conventional theory) and Weyl himself introduced it, not for its own sake, but in order to establish a connection between electromagnetism and gravitation. But, ironically, the gauge principle has turned out to be of more immediate use for the other fundamental interactions, namely, the weak and strong interactions and, as will be seen in section 7.3, Weyl's approach is by far the most natural one for the latter interactions. Meanwhile, it is worth noting that, in contrast to the interaction of matter with itself, where the couplings are arbitrary and the form of the interaction is restricted only by Poincaré invariance and renormalization, the interaction of the EM field with matter is completely determined once the charges of the matter fields are assigned. Thus for the EM field, the symmetry determines the dynamics.

7.3   The gauge principle for simple Lie groups

It is now generally accepted that Weyl's gauge principle can be used to describe the strong and weak interactions as well as the EM ones by generalizing $U(1)$ to other compact Lie groups. The first extension of the principle, to the isospin $SU(2)$ group, was made by Yang and Mills (1954) and was afterwards generalized to other compact simple groups by Utiyama (1956), and by Gell-Mann and Glashow (1961). In this section the generalization to any compact simple group will be described.

The starting point is the matter–matter Lagrangian (6.3) which is assumed to be invariant with respect to the rigid compact simple group G in question,

$$L(\Phi(x), \partial_\mu \Phi(x)) = L(U(g)\,\Phi(x), U(g)\,\partial_\mu \Phi(x)), \qquad (7.8)$$

where $g \in G$, $(\Phi(x) = \phi(x), \psi(x), \chi(x))$. When the rigid group is gauged, $g \to g(x)$, the kinetic terms in (7.8) do not remain invariant. So, in analogy to the EM case, one introduces a set of vector potentials A_μ^k, $k = 1, \ldots, n$, one for each generator σ^k of G, one defines the covariant derivative

$$\mathrm{D}_\mu \Phi = \partial_\mu \Phi + e(A_\mu \cdot \sigma)\,\Phi, \quad \text{or} \quad (\mathrm{D}_\mu \Phi)^a = \partial_\mu \Phi^a + eA_\mu^k \sigma_k^{ab} \Phi^b, \quad (7.9)$$

where $e \neq 0$ is a dimensionless constant, and then one requires that the $A_\mu^k(x)$ transform in such a way that the derivative D_μ will be covariant, i.e. one requires that

$$U(g(x))\,\mathrm{D}_\mu \Phi(x) = \mathrm{D}_\mu U(g(x))\,\Phi(x). \qquad (7.10)$$

Since $\Phi(x)$ is in an arbitrary representation of G, this requirement is equivalent to the condition

$$\mathrm{D}_\mu(A^g) = U(g(x))\,\mathrm{D}_\mu(A)\,U^{-1}(g(x)), \qquad (7.11)$$

where $A_\mu = A_\mu^k \sigma_k$, and on decomposing (7.11) one finds that it is equivalent to the transformation law

$$A_\mu^g = U A_\mu U^{-1} + U \partial_\mu U^{-1}, \tag{7.12}$$

for the vector potential A_μ. A vector potential which transforms in this manner is called a gauge potential. Note that (7.12) is a fusion of the familiar laws

$$A_\mu \to U A_\mu U^{-1}, \quad A_\mu \to A_\mu + \partial_\mu \Lambda, \tag{7.13}$$

for non-abelian rigid and abelian gauge groups respectively. Note also that the infinitesimal version of (7.12) is $\delta A_\mu = D_\mu \Lambda$, where $U = \exp \Lambda$, which is the obvious non-abelian generalization of the abelian $\delta A_\mu = \partial_\mu \Lambda$. Equation (7.12) implies, of course, that A_μ lies in the Lie algebra of the group.

Thanks to the requirement (7.10) the Lagrangian remains invariant when $g \to g(x)$, i.e.

$$L(\Phi(x), D_\mu \Phi(x)) = L(U(g(x)) \Phi(x), U(g(x)) D_\mu \Phi(x)), \tag{7.14}$$

and it remains only to construct a kinetic term for the gauge fields A_μ themselves, similar to the Maxwell term F^2. For this purpose one defines the non-abelian gauge field $F_{\mu\nu}(x)$ according to

$$F_{\mu\nu} = e^{-1}[D_\mu, D_\nu] = \partial_\mu A_\mu - \partial_\nu A_\mu + e[A_\mu, A_\nu]. \tag{7.15}$$

This field is not gauge invariant, as in the EM case, but, since it is constructed from the Ds, it is gauge *covariant*,

$$F_{\mu\nu}(A^g) = U F_{\mu\nu}(A) U^{-1}, \tag{7.16}$$

and this is sufficient for the construction of a kinetic term. In fact the kinetic term is defined to be the trace of F^2 and is thus gauge invariant. The final Lagrangian is therefore

$$L(\Phi, A) = -\tfrac{1}{4} \mathrm{tr}\, F_{\mu\nu} F^{\mu\nu} + L(\Phi, D_\mu \Phi). \tag{7.17}$$

This is the required gauge extension of $L(\Phi, \partial_\mu \Phi)$, and is manifestly invariant with respect to G when $g \to g(x)$. When $L(x)$ in (7.8) is replaced by $L(x)$ in (7.17), L is said to be 'gauged with respect to G'.

Note that the covariance of $F_{\mu\nu}$ is achieved only by adding a nonlinear term (commutator) to the usual curl-term. Apart from this term the expression (7.17) is formally the same as in the EM case. In fact the only distinction between (7.17) and a Lagrangian gauged with r copies of $U(1)$ lies in this term and in the fact that the $U(1)$ charge assignments are not arbitrary but are dictated by the representations of G to which the matter fields belong. Note that the assignment of the matter fields $\phi(x)$, $\psi(x)$, $\chi(x)$

to the representations of G is the generalization of the charge assignments in the EM case, and it is more restrictive because when G is simple the non-trivial CUIRs are multidimensional. Thus the fact that the symmetry determines the dynamics is even more evident in the non-abelian case, because the charge e is determined by the self-interaction of the gauge field, and the dimensions of the CUIRs are fixed. Indeed, the *only* freedom in the non-abelian case is the choice of representation for the matter fields.

The Lagrangian (7.17) contains no mass term $M_{\alpha\beta}A_\mu^\alpha A_\mu^\beta$ for the gauge fields, since this is excluded by invariance with respect to the transformation (7.11). At first sight, this would seem to be a serious drawback because it is known experimentally that there is only one massless gauge field, namely the EM field. This problem is overcome by spontaneous symmetry breaking, as described in the next chapter. In fact, the reason for the presence of scalar fields in (7.17) is to generate the gauge-field masses by a spontaneous breakdown.

There are a few technicalities that might be mentioned at this point. First, it is easily verified that the gauge field satisfies the Bianchi identity

$$\sum_{\text{cyclic}} D_\mu F_{\lambda\sigma} = 0, \quad \text{or} \quad D_\mu \tilde{F}_{\mu\nu} = 0, \qquad (7.18)$$

where $\tilde{F}_{\mu\nu} = \tfrac{1}{2}\epsilon_{\mu\nu\lambda\sigma}F_{\lambda\sigma}$, and field equations

$$D_\mu F_{\mu\nu} = \frac{\delta L}{\delta A_\nu}, \qquad (7.19)$$

analogous to Maxwell's equations. Second, the gauge potential A_μ and the gauge field $F_{\mu\nu}$ may be obtained in component form by using the usual generators σ^a of (5.17) to write

$$A_\mu^a = \text{tr}\,\sigma^a A_\mu, \quad F_{\mu\nu}^a = \text{tr}\,\sigma^a F_{\mu\nu}. \qquad (7.20)$$

Then $F_{\mu\nu}^a$ is seen to take the form

$$F_{\mu\nu}^a = \partial_\mu A_\nu^a - \partial_\nu A_\mu^a + e f_{bc}^a A_\mu^b A_\nu^c = g_{\mu\nu}^a + e f_{bc}^a A_\mu^b A_\nu^c, \qquad (7.21)$$

where f_{bc}^a are the structure constants. The kinetic term for the gauge fields takes the form

$$-L_{\text{kin}}(F) = \tfrac{1}{4} g_{\mu\nu}^a g_{\mu\nu}^a + \tfrac{1}{2} e f_{bc}^a A_\mu^b A_\nu^c g_{\mu\nu}^a + \tfrac{1}{4} e^2 f_{bc}^a f_{de}^a A_\mu^b A_\mu^c A_\nu^d A_\nu^e. \qquad (7.22)$$

Finally, the generalization to an arbitrary compact Lie group $G = U(1) \times U(1) \times \ldots \times U(1) \times G_1 \times G_2 \times \ldots \times G_n \pmod{Z}$ where the G_k are simple is obtained by writing

$$A_\mu = \sum_t e_t \overset{(t)}{A_\mu} \overset{(t)}{\sigma} + \sum_s e_s \overset{(s)}{A_\mu^k} \overset{(s)}{\sigma_k}, \qquad (7.23)$$

where $\overset{(t)}{e}$, $\overset{(s)}{e}$ are independent coupling constants (one for each irreducible part of G) and $\overset{(t)}{\sigma}$ and $\overset{(s)}{\sigma_a}$ are the generators of the $U(1)$s and G_ks respectively. The $\overset{(t)}{\sigma}$ are usually chosen diagonal. An important point to note is that since G is a direct product the gauge fields belonging to the different irreducible parts do not interact directly (through (7.22)) but only through the matter fields. The choice of coupling constants in (7.23) and the assignment of the matter fields to representations of G are the only free parameters for the gauge–gauge and gauge–matter interactions.

Although gauge theory is introduced in the above inductive manner for historical and pedagogical reasons it is clear that the essential ingredients – the gauge potential, the gauge field, and the covariant derivative – have an intrinsic mathematical structure which is independent of the context. This structure has been well studied by mathematicians, in the context of differential geometry. In this context transformations $g(x)$ are identified as sections of principal bundles, with Minkowski space $\mathcal{M}$ as base and the Lie groups G as fibres, the scalar and fermion fields are identified as sections of vector bundles with base $\mathcal{M}$, the gauge potential as a connection form for $G(x)$, and $F_{\mu\nu}(x)$ as the components of the curvature. A review of these aspects is given by Daniel and Viallet (1980). The fibre-bundle formulation is not necessary for dealing with those aspects of gauge theory which are local in $\mathcal{M}$, but it becomes important for understanding problems, such as the axial anomaly (next section) and the gauge-fixing ambiguity (Gribov, 1977; Singer, 1978) which are of global origin. It also shows that gauge theory, and thus the theory of strong, weak and electromagnetic interactions, is basically a geometrical theory. This is not only aesthetically pleasing but brings the unification of weak, electromagnetic and strong interactions with gravitation a step closer.

7.4 Renormalization constraints: the axial anomaly

As mentioned in the introduction, the quantization and renormalization of gauge theories are beyond the scope of this monograph (see suggestions for further reading, e.g. Collins (1984)). However, there are some semi-classical constraints (more precisely, constraints that come from quantum renormalization but can be implemented at the classical level) on the matter field assignments, and since these constraints are important for model building they will be treated briefly in this and the next section.

The first constraint comes from the fact that axial fermion currents, which are conserved at the classical level, are not necessarily conserved after second quantization. More precisely, if $j_\mu^a = \bar{\psi}\gamma_5\gamma_\mu\sigma^a\psi$ is an axial fermion current which is conserved at the classical level ($\partial_\mu j_\mu^a = 0$) then, after second quantization it is no longer conserved but satisfies the equation

$$\partial_\mu j_\mu^a = \frac{e^2}{24\pi^2} T_{bc}^a (f) G^{bc}(A_\mu), \tag{7.24}$$

where the T_{bc}^a, which are symmetric in b and c, are constants that depend only on the group and the fermion representation f (see below) and the $G^{bc}(A_\mu)$ are pseudo-scalars that depend only on the gauge potentials. Explicitly,

$$G^{bc}(A_\mu) = \epsilon_{\mu\nu\alpha\beta}[g_{\mu\nu}^b g_{\alpha\beta}^c + \partial_\mu(A_\nu^b f_{rs}^c A_\alpha^r A_\beta^s)] = \partial_\mu i_\mu^{bc} \tag{7.25}$$

where

$$i_\mu^{bc} = \epsilon_{\mu\nu\alpha\beta} A_\nu^b (2g_{\alpha\beta}^c + f_{rs}^c A_\alpha^r A_\beta^s),$$

the quantities $g_{\mu\nu}^b$ are the curls $\partial_\mu A_\nu^b - \partial_\nu A_\mu^b$ and the f_{rs}^c are the structure constants. The result (7.24) is readily derived in the first order of the loop expansion (see, for example, Huang, 1982) and is actually left unchanged by the higher orders (Bardeen, 1973). The reason that there are no higher loop contributions, and that $G^{bc}(A_\mu)$ can be expressed as a total divergence, is that the currents i_μ^{bc} have a topological significance (Zumino, 1984). Note that, in contrast to the abelian case $G^{bc} = \epsilon_{\mu\nu\alpha\beta} g_{\mu\nu} g_{\alpha\beta}$ (and to the well-known charge-density $G = \epsilon_{\mu\nu\alpha\beta} F_{\mu\nu}^a F_{\alpha\beta}^a$ for instantons) the non-abelian G^{bc} cannot be expressed in terms of the gauge fields $F_{\mu\nu}^a$ alone.

A serious problem that arises from the right-hand side of (7.24) (which is called the ABJ anomaly as it was first discussed by Adler (1969, 1970) and by Bell and Jackiw (1969)) is that, whenever it is not zero it destroys the renormalizability of the theory. This is because the current (J_μ) which is conserved and the current (j_μ) that couples to the gauge fields in the Lagrangian do not coincide (Gross and Jackiw, 1972). Thus, in order to preserve renormalizability one must arrange that the coefficients T_{bc}^a in the anomaly vanish. In other words, in order to construct renormalizable theories, the usual conditions (Yukawa couplings and fourth-degree potentials) are not sufficient, but must be supplemented by the ABJ condition that the T_{bc}^a be zero. This condition is purely algebraic and, in general, it places constraints on the fermion assignments. These constraints can be implemented at the classical level and in practice they are used at this level for model building.

The coefficients T_{bc}^a are easily computed because they come only from the first order in perturbation theory, and, computing them (Huang, 1982) and setting them equal to zero, one obtains

$$T_{bc}^a = L_{bc}^a - R_{bc}^a = 0, \quad (L, R)_{bc}^a = \mathrm{tr}\,\sigma_{L, R}^a \{\sigma_{L, R}^b \sigma_{L, R}^c\} = 2g^{abc} I_3(L, R), \tag{7.26}$$

where $\sigma_{L,R}$ are the generators of the left- and right-handed fermion representations respectively, g^{abc} are the invariant symmetric tensors used to construct the third-degree Casimir (5.35) in section 5.6 and $I_3(L, R)$ are the third-degree indices of (5.37), in section 5.6. The representations σ_L and σ_R need not be irreducible and often are not.

It is clear that the ABJ conditions will be automatically satisfied if the fermion representations are such that $C_3 = 0$, and such representations are called 'safe'. Similarly, groups for which all representations have $C_3 = 0$ are called 'safe' (Georgi and Glashow, 1972). From section 5.6 one sees that this includes all the simple groups except $SU(n)$, E_6 (and $SO(6) \approx SU(4)$). Furthermore, if the left- and right-handed fermion assignments are 'vectorlike', i.e. σ_L and σ_R are equivalent, $\sigma_L^a = U\sigma_R^a U^{-1}$, then evidently $C_3^L = C_3^R$ and the anomaly vanishes. Thus the ABJ condition actually applies only to non-vectorlike assignments in $SU(n)$, E_6 and the $U(1)$ groups. However, because of the parity violation in the weak interactions and the $U(n)$ character of most gauge groups, such assignments are common.

For the $U(1)$ groups T_{bc}^a reduces to $\mathrm{tr}\,\sigma_L^3 - \mathrm{tr}\,\sigma_R^3$ and thus the ABJ condition requires that the sum of the cubes of the $U(1)$ charges for the left- and right-handed fermions be the same. The most important and far reaching consequence of this requirement is in the standard $U(2)$ electroweak model (chapter 9) where the $SU(2)$ subgroup of $U(2)$ is safe, but the $U(1)$ subgroup requires the number of quark and lepton families be equal.

For grand unification the primitive representations F_k of $SU(n)$, $n > 3$ are often used, and for these the indices $I_3(k)$ turn out to be

$$I_3(k) = -I_3(n-k) = c\frac{(n-2k)}{(k-1)!\,(n-k-1)!}, \quad k = 1, ..., n-1, \quad (7.27)$$

where c is a constant (independent of k). The fermion assignments must then be chosen so that $I_3(L) = I_3(R)$, a well-known example being the 5* and 10 of $SU(5)$ for which $I_3 = \pm c$.

7.5 Renormalization constraints: asymptotic freedom

The second constraint on the assignments of the matter fields is not necessary for renormalization as such, but for an important property of the theory that follows from renormalization, namely, 'asymptotic freedom'. This property may be described briefly as follows: Let g be a dimensionless coupling constant for a quantum field theory in four dimensions, and, for simplicity, assume first that g is the only such constant. Then, on account of the renormalization, the effective coupling for fields with momentum of order μ is not g, but $g(\mu)$, where $g(\mu)$ is determined by the so-called renormalization group equation

$$\mu \frac{\mathrm{d}}{\mathrm{d}\mu} g(\mu) = \beta(g(\mu)) = 2bg^2(\mu) + cg^3(\mu) + \dots, \qquad (7.28)$$

in which β is a function of $g(\mu)$ only, and, for sufficiently small $g(\mu)$, has the perturbation expansion shown (for a lucid derivation of (7.28) see Weinberg, 1973). The factor 2 is inserted in (7.28) so that in the case of gauge coupling, for which g is identified as e^2 rather than e, the coefficient b has the conventional value. The most important property of β is its sign (which in perturbation theory reduces to the sign of the first non-vanishing coefficient). If β is negative, the effective coupling constant *decreases* with increasing scale and eventually becomes zero. Then the theory is said to be asymptotically free. Note that if the perturbation expansion is valid and the first non-vanishing coefficient, b say, is negative, the use of perturbation for $\mu \to \infty$ is self-consistent since $g(\mu)$ is decreasing. To order g^2, (7.28) can be integrated, and yields

$$f(\mu) = f(\mu_0) - 2b \ln(\mu/\mu_0), \qquad (7.29)$$

where $f(\mu) = g^{-1}(\mu)$, and μ_0 is the initial point. More generally if there are a number of dimensionless coupling constants g^α, then (7.28) becomes

$$\mu \frac{\mathrm{d}}{\mathrm{d}\mu} g^\alpha = 2b^\alpha_{\beta\gamma} g^\beta(\mu) g^\gamma(\mu) + \dots, \qquad (7.30)$$

and if the matrices $b^\alpha_{\beta\gamma}$ for each α are negative, the theory is said to be asymptotically free.

Until the advent of non-abelian gauge theories, asymptotic freedom was an academic concept, because, for all renormalizable matter–matter and gauge-field–matter interactions the matrices $b^\alpha_{\beta\gamma}$ in (7.30) were positive. But non-abelian gauge-theories are found to have the special property that these coefficients are negative (Politzer, 1974). More precisely, for each gauge field belonging to a simple group G, and not interacting with matter (pure Yang–Mills field), the constant $2b$ in (7.28) is

$$2b = -\tfrac{11}{3}\left(\frac{1}{8\pi^2}\right) C_2(G), \qquad (7.31)$$

where $C_2(G)$ is the Casimir operator for the adjoint representation (and g in (7.28) is identified as the *square* of the gauge coupling constant

$(g = e^2)$). More generally, for a gauge field interacting with matter the constant $2b$ turns out to be

$$2b = \frac{1}{8\pi^2}[-\tfrac{11}{3}I_2(G)+\tfrac{4}{3}fI_2(f)+\tfrac{1}{6}sI_2(s)], \quad I_2(j) = \frac{\dim(j)}{\dim(adj)}C_2(j), \quad (7.32)$$

where $I(f)$ and $I(s)$ denote the indices of the irreducible fermion and scalar representations, as already discussed in section 5.5, and f and s denote the number of times each such representation occurs. Thus the non-abelian gauge couplings are asymptotically free provided the number of matter fields with which they interact is not too large, the precise limitation being given by (7.32).

So long as the gauge fields interact only with fermions, the only couplings are the gauge couplings (the same for gauge–gauge and gauge–fermion) and so (7.32) with $I(s) = 0$ is the whole story. However, if scalar fields are present, one must consider also the Yukawa couplings and the couplings in the scalar potential $V(\phi)$. Since these couplings are not asymptotically free in the absence of non-abelian gauge fields they can only become free if the gauge coupling reverses the usual sign, and as one might expect, this can happen only under special conditions. For the Yukawa couplings G, the conditions are not too restrictive, since the general form of the one-loop renormalization equation for the Yukawa constant is

$$\mu\frac{dG}{d\mu} = \alpha G^3 - \beta Ge^2 \quad \text{or} \quad \mu\frac{dF}{d\mu} = e^2F[\alpha F^2 - \beta + |b|], \quad (7.33)$$

where $F = (G/e)$, α is positive as in gauge field independent theories, and β turns out to be positive for the same reason that $(-2b)$ is positive in (7.31). From (7.33) one sees that a sufficient condition for the (one-loop) asymptotic freedom of G is $\beta > |b|$, and then it is free in the sector $(G(g) < (\beta-|b|)/\alpha)$.

For the scalars the conditions are much more stringent. The renormalization group equations for a single self-interaction coupling constant f, with gauge coupling constant e, is of the general form

$$\mu\frac{d}{d\mu}f = Af^2 - Bfe^2 + Ce^4 \quad \text{or} \quad \mu\frac{d}{d\mu}h = e^2[Ah^2 - (B-2|b|)h+C], \quad (7.34)$$

where $h = (f/e^2)$, and A, B, C are numerical coefficients depending on the group and representation, and which in general are positive (Politzer, 1974). For example, for the fundamental representation of $SU(n)$,

$$A = n+4, \quad B = 3(n^2-1)/n, \quad C = 3(n-1)(n^2+2n-2)/4n^2. \quad (7.35)$$

It is clear from (7.34) that a sufficient condition for asymptotic freedom is

$$(B - 2|b|) > (4AC)^{\frac{1}{2}} > 0 \qquad (7.36)$$

and if this condition is satisfied h tends to the fixed point ρ, where

$$2A\rho = (B - 2|b|) \pm [(B - 2|b|)^2 - 4AC]^{\frac{1}{2}} \qquad (7.37)$$

and f tends to zero like e^2. The inequality (7.36) can be satisfied, but only in very special cases, notably when the group is large and the scalar representation is small (Gross and Wilczek, 1973). Unfortunately, it is very difficult to reconcile this condition with the requirement that the scalar fields generate a non-trivial spontaneous breakdown. An interesting exception to the rule occurs in the supersymmetric case, in which the scalar and Yukawa couplings coincide and so the parameter A, which is normally positive-definite, vanishes (Fayet and Ferrarra, 1977).

The case of a single coupling constant as in (7.34) is, of course, only the simplest case, since, in general there are many scalar couplings h, and Yukawa couplings G, to be taken into account. However, (7.34) is a good example because even in the most general case the right-hand side of the (one-loop) renormalization group equation is a homogeneous quadratic form (in h, G^2 and e^2).

The scalar conditions (7.36) do not cause any great difficulty for the standard $S(U(3) \times U(2))$ theory of the strong and electroweak interactions, because the only part which is asymptotically free (the strong, or coloured, part discussed in the next chapter) is assumed to be unbroken, and hence does not contain any scalars. However, the inequalities do create problems for grand-unified theories (chapter 10).

Exercises

7.1. Gauge the $SU(2)$-invariant Lagrangian (6.4).
7.2. Show that for real or pseudo-real unitary representations the generators satisfy the reality condition $\sigma^* = -W\sigma W^{-1}$, and deduce that for such representations the anomaly (7.26) vanishes.
7.3. Show that for $SU(n)$ the maximum number of fermion multiplets consistent with asymptotic freedom is 2 for the adjoint, and $11n/2$ for the fundamental, representations.

8

Spontaneous symmetry breaking

8.1 Motivation and definition

As discussed in chapter 7 the Yang–Mills–Higgs Lagrangian density (7.17) contains no explicit mass terms (i.e. terms of the form $m_{ab}A_\mu^a A_\mu^b$) for the gauge fields. This is a serious problem because it is known experimentally (from the Meissner effect in superconductivity and the existence of the $W_\mu^\pm$ and Z_μ^0 in weak interactions) that such terms exist. It cannot be solved by simply inserting the required mass terms by hand, because *ad hoc* mass terms destroy the gauge invariance, and hence the renormalizability, of the theory. But it can be solved by inducing the mass terms through a spontaneous symmetry breakdown (SSB), and this is the motivation for SSB.

In the context of Lagrangians with scalar potentials, SSB is defined as follows: Suppose that the scalar fields $\phi(x)$ are assigned to a continuous unitary (not necessarily irreducible) representation (CUR) $U(g)$ of the rigid internal group G and that the potential is group invariant, i.e.

$$V(U(g)\phi) = V(\phi), \tag{8.1}$$

where $V(\phi)$ is the potential. Then if the point $\overset{\circ}{\phi}$ where $V(\phi)$ takes its minimum is not G invariant,

$$U(g)\overset{\circ}{\phi} \neq \overset{\circ}{\phi} \quad \text{for some } g \in G, \tag{8.2}$$

the symmetry is said to be spontaneously broken. The classic example is

$$V = \lambda(\phi^2 - c^2)^2. \tag{8.3}$$

where λ and c are constants and ϕ belongs to the n-dimensional representation of $SO(n)$. It is clear that in this case the minimum of V does not occur at the symmetry point $\phi = 0$ but at one of the non-symmetrical points $\overset{\circ}{\phi} = c\overset{\circ}{u}$, where $\overset{\circ}{u}$ is any unit vector. (For most potentials $\overset{\circ}{\phi} = 0$ is the only symmetry point, so if the minimum occurs for $\overset{\circ}{\phi} \neq 0$ there is a spontaneous breakdown.)

Although the minimum point $\overset{\circ}{\phi}$ for a SSB is not invariant for all $g \in G$

there may, of course, be *some* elements $g \in G$ for which it is invariant. It is easy to see that all such elements form a (closed) subgroup of G and this subgroup, which will be denoted by H, is called variously the little group of $\overset{\circ}{\phi}$, the stability group of $\overset{\circ}{\phi}$, the isotropy group of $\overset{\circ}{\phi}$, and the residual symmetry group of G. Formally, it is the group H defined by

$$U(h)\,\overset{\circ}{\phi} = \overset{\circ}{\phi} \Leftrightarrow h \in H. \tag{8.4}$$

In the $G = SO(n)$ example $H = SO(n-1)$. Note that the concept may be extended to include the unbroken case by including $H = G$.

It is easy to see from (8.1) that when a SSB takes place the minimum point $\overset{\circ}{\phi}$ is not unique. In fact any point on the orbit $U(g)\overset{\circ}{\phi}$ (which is just the coset space G/H) will be a minimum point. Thus an alternative way to define a SSB is to say that the potential minimum is degenerate (its orbit consists of more than one point).

Finally, one sees that since

$$U(h)\,\overset{\circ}{\phi} = \overset{\circ}{\phi} \Rightarrow U(ghg^{-1})\,U(g)\,\overset{\circ}{\phi} = U(g)\overset{\circ}{\phi}, \tag{8.5}$$

the little group is the same (up to conjugation) for all points on the orbit. Collecting all these results, one sees that the natural, group-invariant, way to characterize a SSB is to specify the *orbit* of the potential minimum, rather than a particular point $\overset{\circ}{\phi}$ on the orbit. In fact only the orbit has a physical meaning. The orbital structure will be considered in much greater detail in chapter 11.

The importance of the minimum $\overset{\circ}{\phi}$ can be seen in a somewhat broader perspective by considering the Lagrangian (7.17) in the Euclidean space formulation. Since the minimizing equations for L are $F_{\mu\nu} = 0$, $D_\mu \phi = 0$, V minimal, and these are gauge equivalent to the equations $A_\mu = 0$, $\partial_\mu \phi = 0$, V minimal (because Euclidean space is topologically trivial) one sees that if $\overset{\circ}{\phi}$ minimizes V, and is *constant*, then it minimizes the whole Euclidean action. Since perturbative QCD is nothing but an expansion about the minimum of the action, one then sees that for $\overset{\circ}{\phi} \neq 0$ the quantum field theory Lagrangian is

$$L(A_\mu, \psi, \theta) = L(A_\mu, \psi, \phi) - L(A_\mu, \psi, \overset{\circ}{\phi}), \tag{8.6}$$

where $\theta(x) = \phi(x) - \overset{\circ}{\phi}$ is the 'true' scalar field (minimizes the potential at $\theta = 0$) and (8.6) is a 'true' Lagrangian (commences with positive quadratic terms). It is clear that $L(A_\mu, \psi, \theta)$ is symmetric only with respect to the little group H, and this is why the symmetry is said to be 'broken'. The remainder of this chapter will be concerned with the detailed structure of $L(A_\mu, \psi, \theta)$.

8.2   Gauge field sector of $L(A_\mu, \psi, \theta)$

From (7.17) and (6.3) one sees that in the case of a SSB the gauge-field part of the Lagrangian $L(A_\mu, \psi, \phi)$ becomes

$$L(A, \psi, \theta) = L_A + L_{A\psi} + \tfrac{1}{2}(D_\mu \theta)^2 + \tfrac{1}{2}(D_\mu \overset{\circ}{\phi})^2 + (D_\mu \theta, D_\mu \overset{\circ}{\phi}), \qquad (8.7)$$

where L_A and $L_{A\psi}$ are as before. Thus for the gauge potential the SSB has induced the last two terms in (8.7).

Since $\overset{\circ}{\phi}$ is constant, the first of these two terms is just a mass for the gauge fields, namely

$$\tfrac{1}{2}(D_\mu \overset{\circ}{\phi})^2 = \tfrac{1}{2}e^2(A_\mu \overset{\circ}{\phi})^2 = \tfrac{1}{2}M_{ab} A_\mu^a A_\mu^b, \quad M_{ab} = e^2(\sigma_a \overset{\circ}{\phi}, \sigma_b \overset{\circ}{\phi}), \qquad (8.8)$$

where σ_a are the generators of the ϕ-representation and e is the gauge coupling constant. (It is assumed for simplicity that the gauge group is irreducible, i.e simple or $U(1)$, but the generalization to the other cases is obvious.) Equation (8.8) shows that masses for the gauge fields are automatically induced by a SSB and they take a very definite form. In particular, since $\sigma_a \overset{\circ}{\phi} = 0$ if, and only if, σ_a is a generator of the little group H, one sees that a gauge field will remain massless if, and only if, it belongs to the residual symmetry group.

One might ask, of course, why the masses in (8.8) are any better than those which would be introduced by hand. The reason is that they take the very special form (8.8) and that they do not come alone, but are accompanied by other induced terms in the Lagrangian such as the last term in (8.7). In fact the induced terms are just those that are necessary to make $L(A_\mu, \psi, \theta)$ invariant with respect to the formal gauge transformation which consists of the usual one for A_μ and ψ but is $\theta \to U(g)\theta + (U(g) - 1)\overset{\circ}{\phi}$ for θ. The important point is that this formal invariance is enough to preserve renormalizability. That is, renormalizability for the unbroken theory implies renormalizability for the broken theory, provided the breaking is spontaneous. Full details can be found in the literature on unified gauge theory given at the end of the monograph.

Note that the sum of the squares of the gauge-field masses in (8.8) is just

$$M_{aa} = e^2 |\overset{\circ}{\phi}|^2 C_2, \qquad (8.9)$$

where C_2 is the average value of the second-order Casimir operator in the ϕ-representation, a result that is useful for estimating the order of magnitude of the gauge-field masses. Note also that since the broken

Lagrangian is still H-invariant, the gauge-field masses must come in H-multiplets, and thus, in general, there are not many different masses.

For any pair (G, H) it is natural to choose the Cartan algebras so that the Cartan algebra of H is a subalgebra of that of G. In many cases (e.g. the defining or adjoint representation) one even has

$$H_i \overset{\circ}{\phi} = \lambda_i \overset{\circ}{\phi}, \quad i = 1, ..., l, \tag{8.10}$$

where H_i form the Cartan algebra of G, and in that case

$$M_{\alpha\beta} = e^2(\overset{\circ}{\phi}, E_{-\alpha} E_\beta \overset{\circ}{\phi}) = e^2(\overset{\circ}{\phi}, [H_j, [E_{-\alpha}, E_\beta]] \overset{\circ}{\phi})(\alpha_j - \beta_j)^{-1} = 0, \tag{8.11}$$

for any two roots with at least one component (j say) unequal, while

$$M_{\alpha\alpha} = e^2(E_\alpha \overset{\circ}{\phi})^2, \quad M_{ij} = e^2 |\overset{\circ}{\phi}|^2 \lambda_i \lambda_j. \tag{8.12}$$

Thus for (8.10) the non-Cartan mass matrix is diagonal, and the Cartan mass matrix is non-zero for at most one field. If $\lambda_i = 0$ (as happens for $\overset{\circ}{\phi}$ in the adjoint representation) all the Cartan masses are zero.

Let us finally consider the last term in (8.7). Since $\overset{\circ}{\phi}$ is constant this term may be written as

$$e(A_\mu \overset{\circ}{\phi}, D_\mu \theta) = e^2(\sigma_a \sigma_b \overset{\circ}{\phi}, \theta) A_\mu^a A_\mu^b - e(\sigma_a \overset{\circ}{\phi}, \theta) \partial_\mu A_\mu^a, \tag{8.13}$$

where an overall divergence has been subtracted. The second term in (8.13) is of little physical importance, and vanishes in two common gauges, namely, the Landau gauge $\partial_\mu A_\mu = 0$ and the physical gauge $(\sigma_a \overset{\circ}{\phi}, \theta) = 0$ (section 8.5). The first term is a non-derivative coupling θA^2 similar to that occurring in scalar quantum electrodynamics (QED). It vanishes for the massless gauge fields (for which $\sigma_a \overset{\circ}{\phi} = 0$) and for the scalar fields in the directions $\sigma_a \overset{\circ}{\phi}$ (i.e. the Goldstone fields of section 8.4) because $(\overset{\circ}{\phi}, \sigma_a \sigma_b \sigma_c \overset{\circ}{\phi})$ is zero. For the scalar field π in the direction $\overset{\circ}{\phi}$ (the polar field of section 8.4) the trilinear term reduces to $(2M_{ab} A_\mu^a A_\mu^b) \pi(x)$ from which one sees that the coupling to π is proportional to the square of the mass of the gauge field.

8.3 Scalar–Fermion sector

The total Dirac sector in the Lagrangian takes the form

$$L(\psi, \phi) = \bar{\psi}_\alpha \overset{}{\not{D}} \psi_\alpha + \bar{\psi}_\alpha M_{\alpha\beta} \psi_\beta + G_{\alpha\beta}^a \bar{\psi}_\alpha \psi_\beta \phi_a + \text{herm. conj.} \tag{8.14}$$

where, for brevity, the scalars and pseudo-scalars are not distinguished, the fermion and scalar representations are not necessarily irreducible, $M_{\alpha\beta}$ is a group-invariant mass matrix and $G_{\alpha\beta}^a$ a set of group-invariant Yukawa couplings. Since (8.14) is linear in the scalar field, the change due to a SSB is very simple, namely,

$$L(\psi, \phi) \to L(\psi, \theta) = \Sigma \, \bar{\psi}_\alpha \, \slashed{D} \psi_\alpha + \bar{\psi}_\alpha (M_{\alpha\beta} + G^a_{\alpha\beta} \overset{\circ}{\phi}_a) \, \psi_\beta$$

$$+ G^a_{\alpha\beta} \bar{\psi}_\alpha \, \psi_\beta \, \theta_a + \text{herm. conj.} \quad (8.15)$$

In other words the only change is that $M_{\alpha\beta} \to M_{\alpha\beta} + G^a_{\alpha\beta} \overset{\circ}{\phi}_a$. Thus the effect of the SSB on the fermions is to split the G-invariant mass multiplets into smaller H-invariant multiplets, (and , if $M_{\alpha\beta} = 0$, to generate H-invariant mass multiplets).

For future use in grand unification theory, it is worth noting that if ψ is split into its chiral (Weyl) components ψ_L and ψ_R, corresponding to $(1 \pm \gamma_5) \, \psi$, then, in contrast to the gauge field which preserves chirality, the Yukawa and mass terms couple L to R. Thus the Weyl decomposition of (8.15) is

$$\bar{\psi} \slashed{D} \psi + \bar{\psi}(m + g\phi)\psi = \bar{\psi}_L \, \slashed{D} \psi_L + \bar{\psi}_R \, \slashed{D} \, \psi_R + \bar{\psi}_L(m + g\phi)\psi_R$$

$$+ \bar{\psi}_R(m + g\phi)^* \, \psi_L. \quad (8.16)$$

In particular, if the left-handed fermions and anti-fermions $((\bar{\psi})_L \text{ not } \overline{(\psi_L)})$ are assigned to a representation f of an internal symmetry group, the antifields must belong to the complex conjugate representation f^*, and then, so far as the internal group is concerned, the coupling (8.16) is

$$f^* \slashed{D} f + f(m + g\phi)f + f^*(m + g\phi)^* f^*. \quad (8.17)$$

Thus D couples to $f \times f^*$, whereas $(m + g\phi)$ couples to $(f \times f)$ and $(f^* \times f^*)$.

8.4 The Goldstone theorem and the scalar potential

In the broken Lagrangian, the scalar potential takes the form

$$V(\theta) = V(\phi) - V(\overset{\circ}{\phi}) = \tfrac{1}{2} V_{\alpha\beta} \, \theta_\alpha \, \theta_\beta + \frac{1}{3!} V_{\alpha\beta\gamma} \, \theta_\alpha \, \theta_\beta \, \theta_\gamma + \frac{1}{4!} V_{\alpha\beta\gamma\delta} \, \theta_\alpha \, \theta_\beta \, \theta_\gamma \, \theta_\delta,$$

plus higher degree terms, where

$$V_{\alpha\beta} = \left(\frac{\partial^2 V}{\partial \phi_\alpha \, \partial \phi_\beta} \right)_{\phi = \overset{\circ}{\phi}}, \text{ etc.} \quad (8.18)$$

In particular, if the potential is renormalizable (fourth degree) the expansion terminates at the quartic term and the latter has the same coefficient as in the original potential $V(\phi)$. For example, for the $SO(n)$-potential of section 8.1,

$$V(\theta) = \frac{\lambda^2 c^2}{3!} \pi^2 + \frac{\lambda^2 c}{3!} \theta^2 \pi + \frac{\lambda^2}{4!} \theta^4, \quad (8.19)$$

where $\pi = (\theta, \dot{u})$. The matrix $V_{\alpha\beta}$ is evidently the mass matrix for the scalar fields, and because $\overset{\circ}{\phi}$ is a minimum of $V(\phi)$, $V_{\alpha\beta}$ must be positive. One of the most important results of SSB is that, for continuous symmetries, $V_{\alpha\beta}$ cannot be positive-definite. The result that $V_{\alpha\beta}$ must have some zero eigenvalues is known as the Goldstone theorem. It shows that a continuous SSB always produces massless scalar fields, as illustrated by (8.19) where all the θ-fields except π are massless. As will be seen shortly, there are always at least $\dim G/H$ massless scalar fields. Here the proof of the Goldstone theorem for classical fields will be given. The extension to quantum field theory may be found in (Goldstone, Salam and Weinberg, 1962; Amit, 1978, 1986).

To establish the theorem for classical fields one considers the symmetry equation (8.1) for infinitesimal $U(g)$, i.e.

$$U(g)\phi - \phi = \delta\phi = (\epsilon_r \sigma_r)\phi + \tfrac{1}{2}(\epsilon_r \epsilon_s \sigma_r \sigma_s)\phi + \dots \qquad (8.20)$$

where ϵ_r are the infinitesimal parameters. One sees at once that

$$\frac{\partial V}{\partial \phi_\alpha}\delta\phi_\alpha = 0, \quad \frac{\partial^2 V}{\partial\phi_\alpha \partial\phi_\beta}\delta\phi_\alpha \delta\phi_\beta = 0, \dots. \qquad (8.21)$$

On setting $\phi = \overset{\circ}{\phi}$, the first equation in (8.21) becomes an identity and the second becomes
$$V_{\alpha\beta}(\sigma_r \overset{\circ}{\phi})_\alpha (\sigma_s \overset{\circ}{\phi})_\beta = 0 \Rightarrow V_{\alpha\beta}(\sigma_s \overset{\circ}{\phi})_\beta = 0, \qquad (8.22)$$

where the implication follows from the fact that $V_{\alpha\beta}$ is real, symmetric and *positive*. Equation (8.22) shows that the vectors $(\sigma_s \overset{\circ}{\phi})_\alpha$ are eigenvectors of $V_{\alpha\beta}$ with eigenvalue zero and hence that $V_{\alpha\beta}$ has at least as many zero eigenvalues as there are linearly independent vectors $(\sigma_r\overset{\circ}{\phi})_\alpha$. But if H is the little group $(\sigma_r\overset{\circ}{\phi})$ is zero if, and only if, σ_r is a generator of H. Thus there are exactly $\dim G/H$ linearly independent non-zero vectors $(\sigma_r\overset{\circ}{\phi})$, as required. Note that the theorem holds for local, as well as absolute, minima, since only the vanishing of V_α and the positivity of $V_{\alpha\beta}$ was used. Note also that the number of Goldstone fields depends only on the dimension of the groups G and H and not on the dimension of the ϕ-representation, a result that will be important for the Higgs mechanism (sections 8.5 and 8.6). A simple geometrical interpretation of the Goldstone theorem is that the Goldstone directions $\sigma_r\overset{\circ}{\phi}$ are just the tangents to the orbit $U(g)\overset{\circ}{\phi}$ along which $V(\phi)$ remains constant.

It may happen, of course, that $V_{\alpha\beta}$ has other fortuitous, zero eigenvalues and then the number of massless scalars is larger than $\dim G/H$. This can happen for example when the invariance group $G(V)$ of the potential is larger than that of the Lagrangian, so that the number of massless fields is dim $G(V)/H > \dim G/H$. It also happens when the parameters in the potential take the critical values for a second-order (smooth in ϕ) phase transition from one little group to another. For example, for ϕ in the fundamental representation of $SO(n)$, and potential $\mu(\phi, \phi) + \lambda(\phi, \phi)^2$, the point $\mu = 0$ is the critical point for a transition from $H = SO(n)$ to $H = SO(n-1)$, and one sees that at this point all n fields in ϕ become massless.

To compute the non-zero scalar masses, more detailed information concerning the groups (G, H), the ϕ-representation, and the potential is needed. However, there are a few general statements that can be made. First, since $V(\theta)$ is H-invariant, the remaining fields must form H-multiplets and thus the number of distinct masses will not, in general, be large. Among the mass multiplets there is always at least one singlet, namely the scalar field in the direction $\overset{\circ}{\phi}$ of the symmetry breakdown. This field, which will be called the polar field $\pi(x)$, is never a Goldstone field, since $(\overset{\circ}{\phi}, \sigma_s \overset{\circ}{\phi}) = 0$, and since the mass matrix is H-invariant, it can 'mix' only with other singlets. In particular, if $\pi(x)$ is the only H-singlet in the representation (as often happens, especially when H is maximal) then $\pi(x)$ has a definite mass squared.

The general fourth-degree potential may be written as

$$V = \tfrac{1}{2} m_{\alpha\beta} \phi_\alpha \phi_\beta + \frac{1}{3!} f_{\alpha\beta\gamma} \phi_\alpha \phi_\beta \phi_\gamma + \frac{1}{4!} g_{\alpha\beta\gamma\delta} \phi_\alpha \phi_\beta \phi_\gamma \phi_\delta, \qquad (8.23)$$

where $f_{\alpha\beta\beta} = 0$, because compact groups have no first-degree invariants, and if the representation is irreducible

$$m_{\alpha\beta} = \delta_{\alpha\beta} m, \quad g_{\alpha\beta\gamma\epsilon} = \tilde{g}_{\alpha\beta\gamma\epsilon} + g(\delta_{\alpha\beta}\delta_{\gamma\epsilon} + \delta_{\alpha\gamma}\delta_{\beta\epsilon} + \delta_{\alpha\epsilon}\delta_{\beta\gamma}) \qquad (8.24)$$

where $\tilde{g}_{\alpha\alpha\gamma\epsilon} = 0$. The extremal equation is evidently

$$\left(\frac{\partial V}{\partial \phi_\alpha}\right)_{\phi=\overset{\circ}{\phi}} = \frac{1}{3!} g_{\alpha\beta\gamma\delta} \overset{\circ}{\phi}_\beta \overset{\circ}{\phi}_\gamma \overset{\circ}{\phi}_\delta + \tfrac{1}{2} f_{\alpha\beta\gamma} \overset{\circ}{\phi}_\beta \overset{\circ}{\phi}_\gamma + m_{\alpha\beta} \overset{\circ}{\phi}_\beta = 0, \qquad (8.25)$$

and the mass matrix is

$$M_{\alpha\beta} = \left(\frac{\partial^2 V}{\partial \phi_\alpha \partial \phi_\beta}\right)_{\phi=\overset{\circ}{\phi}} = m_{\alpha\beta} + f_{\alpha\beta\gamma} \overset{\circ}{\phi}_\gamma + \tfrac{1}{2} g_{\alpha\beta\gamma\delta} \overset{\circ}{\phi}_\gamma \overset{\circ}{\phi}_\delta. \qquad (8.26)$$

In particular the sum of the squares of the masses in the irreducible case is just $dm + \tfrac{1}{2} g (d+2)(\overset{\circ}{\phi}, \overset{\circ}{\phi})$ where d is the dimension of the representation.

It is easy to see that variation of V with respect to the polar field π is equivalent to variation with respect to the scale s of the fields ϕ ($s^2 = (\phi, \phi)$,

$s(\partial/\partial s) = \phi_\alpha(\partial/\partial\phi_\alpha) = \pi(\partial/\partial\pi))$ and if V is decomposed into its parts of different degree,

$$V = \frac{V_2}{2} + \frac{V_3}{3!} + \frac{V_4}{4!}, \qquad (8.27)$$

then the scale variation yields

$$s\frac{\partial V}{\partial s} = V_2 + \frac{V_3}{2} + \frac{V_4}{6} = 0, \quad s^2\frac{\partial^2 V}{\partial s^2} = s^2\langle\pi|M|\pi\rangle = V_2 + V_3 + \tfrac{1}{2}V_4, \quad (8.28)$$

where $\langle\pi|M|\pi\rangle$ is the expectation value of the mass matrix in the π-direction. (Note that these results can also be obtained by taking the inner product of (8.25) and (8.26) with $\dot{\phi}$.)

An interesting and useful consequence of (8.28) is that, if $V_3 = 0$ (V is reflexion invariant) then $\langle\pi|M|\pi\rangle$ is given by the formula

$$\langle\pi|M|\pi\rangle = -2V_2(\dot{\phi})/(\dot{\phi},\dot{\phi}), \qquad (8.29)$$

which is universal in the sense that it holds for any group G, any representation (reducible or not) and any fourth-degree reflexion invariant potential. Thus it allows $\langle\pi|M|\pi\rangle$ to be obtained from the potential by inspection. In particular, in the irreducible case, one has $\langle\pi|M|\pi\rangle = -2m$ where m is the mass parameter in (8.24).

Finally, it is interesting to note that if the polar field has a definite mass squared, m_π^2 say, then $\dot{\phi}$ is an eigenvector of the mass matrix $M_{\alpha\beta}$ with this eigenvalue, and the cubic extremal equation reduces to the quadratic equation

$$f_{\alpha\beta\gamma}\dot{\phi}_\beta\dot{\phi}_\gamma + 4m_{\alpha\beta}\dot{\phi}_\beta + 2m_\pi^2\dot{\phi}_\gamma = 0. \qquad (8.30)$$

In general, (8.30) is a much more severe limitation on $\dot{\phi}$ than is (8.25).

8.5 Higgs mechanism

Although the spontaneous symmetry breaking solves the problem of the gauge-field masses, it introduces two further problems, one experimental and one theoretical. The experimental problem is the appearance of the Goldstone fields. Because they are massless such fields are long range and should be observed. But no long-range scalars have been seen experimentally. The theoretical problem is most easily seen by considering the canonical commutation relations

$$[A_\mu(x), A_\nu(y)] = g_{\mu\nu}\,\mathrm{D}(x-y), \qquad (8.31)$$

for the gauge fields, where $\mathrm{D}(x-y)$ is the usual massless c-number D-function (Bogolibov and Shirkov, 1954). As is well known, the indefiniteness of the Minkowski metric $g_{\mu\nu}$ leads to a negative norm for the time component A_0 of A_μ. For massless fields this problem is resolved by the cancellation of the A_0 contribution with the longitudinal spacelike component $\mathbf{p} \cdot \mathbf{A}$, leaving two physical transverse components $\mathbf{p} \times \mathbf{A}$, as befits a massless radiation field (Gupta–Bleuler mechanism). But for massive fields the problem cannot be solved this way because all three spacelike components are genuine physical fields.

It took some time (1959–67) for physicists to realize that these two problems actually cancel each other, i.e. that there is a generalized Gupta–Bleuler mechanism by which the Goldstone scalars and the timelike component of the gauge field (more precisely $\partial_\mu A_\mu$) cancel, leaving the three spacelike components of A_μ intact. This mechanism is known as the Higgs mechanism, and will now be described.

Let us begin, for simplicity, with the abelian case, and a scalar–QED Lagrangian density of the form

$$-L(x) = \tfrac{1}{2}(E^2 - B^2) + \tfrac{1}{2}|\partial_\mu + ieA_\mu \phi|^2 + \lambda(|\phi|^2 - d)^2, \qquad (8.32)$$

where ϕ is a single complex field. This is actually the Lagrangian for the Landau–Ginsburg model of superconductivity (Fetter and Walecka, 1971). For $d = c^2 > 0$ there is a SSB (at $\overset{\circ}{\phi}_1 = c$, $\overset{\circ}{\phi}_2 = 0$ say) and the Lagrangian becomes

$$-L(A_\mu, \theta) = \tfrac{1}{2}(E^2 - B^2) + \tfrac{1}{2}|\partial_\mu + ieA_\mu \theta|^2 + \tfrac{1}{2}e^2c^2A_\mu^2$$
$$+ e^2c\theta_1 A_\mu^2 - ec\theta_2(\partial_\mu A_\mu) + \lambda[4c^2\theta_1^2 + 4c\theta_1|\theta|^2 + |\theta|^4], \quad (8.33)$$

in accordance with the general results of the previous sections. Clearly the field θ_2 is the Goldstone field.

The SSB in (8.33) was made in an arbitrary gauge. Suppose, however, that before making it one had chosen the gauge so that ϕ was real (which is certainly possible since the gauge group $U(1)$ is just the phase group of ϕ). Then the Lagrangian (8.32) would reduce to

$$-L(A, \rho) = \tfrac{1}{2}(E^2 - B^2) + \tfrac{1}{2}(\partial_\mu \rho)^2 + \tfrac{1}{2}e^2A_\mu^2\rho^2 + \lambda(\rho^2 - c^2)^2, \qquad (8.34)$$

and on breaking it, i.e. by writing $\rho = \sigma + c$ one would obtain

$$-L(A, \sigma) = \tfrac{1}{2}(E^2 - B^2) + \tfrac{1}{2}(\partial_\mu \sigma)^2 + \tfrac{1}{2}e^2c^2A_\mu^2$$
$$+ e^2c\sigma A_\mu^2 + \tfrac{1}{2}e^2A_\mu^2\sigma^2 + \lambda[4c^2\sigma^2 + 4c\sigma^3 + \sigma^4], \quad (8.35)$$

which is just (8.33) without the Goldstone field. Thus by choosing a suitable gauge the Goldstone field is eliminated.

Once the gauge has been fixed for ϕ, however, it is no longer free for

A_μ, and to investigate the form of A_μ in this gauge, it is convenient to consider the electromagnetic current

$$j_\mu = e\phi^* \overset{\leftrightarrow}{\partial}_\mu \phi + e^2 A_\mu |\phi|^2, \tag{8.36}$$

which, after the SSB ($\phi \to \theta + c$) becomes

$$j_\mu = ec\partial_\mu \theta_2 + e^2 c^2 A_\mu + \text{nonlinear terms.} \tag{8.37}$$

The current conservation law $\partial^\mu j_\mu = 0$ then gives

$$\Box \, ec\theta_2 + e^2 c^2 (\partial_\mu A_\mu) = \text{nonlinear terms.} \tag{8.38}$$

It follows that for the asymptotic (in and out) fields

$$\Box \, \theta_2 = -ec(\partial_\mu A_\mu), \tag{8.39}$$

and thus in the gauge in which $\theta_2 = 0$, the component $\partial_\mu A_\mu$ of the free gauge field also vanishes. But since the gauge field is massive after the SSB the component $\partial_\mu A_\mu$ is just $p^\mu A_\mu$, which, in the rest-frame, is just A_0. Hence when $\theta_2 = 0$, $A_0 = 0$ also. Conversely, when $A_0 = 0$ then $\Box \theta_2 = 0$ so θ_2 decouples (and with suitable boundary conditions it vanishes).

These results show that there exists a gauge in which A_0 and θ_2 simultaneously vanish, and this is the Higgs mechanism for the abelian case. Because the remaining scalar field θ_1 (which is just the polar field for this case) and the three spacelike components of A_μ are all (massive) physical fields, the gauge in which A_0 and θ_2 vanish is called the *physical* gauge. Because there are no fields of negative norm, it is also called the *unitary* gauge.

The generalization of the Higgs mechanism to the non-abelian case is not completely trivial because the gauge transformation which eliminates the Goldstone fields can no longer be written in an explicit form, but only shown to exist. The existence proof is as follows: consider the inner product

$$f(a) = (\phi(x), U(g)\overset{\circ}{\phi}), \quad g = \exp a \cdot \sigma, \tag{8.40}$$

where $\phi(x)$ is the scalar field in a given gauge and $\overset{\circ}{\phi}$ the constant point of the potential minimum. For each fixed value of x, $f(a)$ may be considered as a function over the group G, and as such is a smooth (even real analytic) function of a, and is bounded (by $|\phi(x)| \, |\overset{\circ}{\phi}|$). Since G is compact, $f(a)$ must therefore take maximum and minimum values on G, i.e. there must exist at least two points $\tilde{a}$ such that

$$\frac{\partial f(\tilde{a})}{\partial \tilde{a}_k} = \left(\phi(x), \frac{\partial U(\tilde{g})}{\partial \tilde{a}_k} \overset{\circ}{\phi} \right) = 0. \tag{8.41}$$

Of course, the point $\tilde{a}$ may be different for different x, so $\tilde{a} = \tilde{a}(x)$. The derivative of $g(a)$ with respect to a is given in (2.17) and applying that result to the representation $U(g)$ one has

$$\frac{\partial U(g(a))}{\partial a_k} = U(g(a))\,\sigma_s\,v_k^a(a). \tag{8.42}$$

Since $v(a)$ is non-singular one therefore has

$$(U^{-1}(\tilde{g})\,\phi(x),\sigma_s\,\mathring{\phi}) = 0, \tag{8.43}$$

and since $\sigma_s\mathring{\phi}$ are exactly the Goldstone directions, this equation shows that the scalar fields $U(\tilde{g}(x))^{-1}\phi(x)$ have no Goldstone components. Thus the gauge transformation $U^{-1}(\tilde{g}(x))$, where $\tilde{g}(x)$ is a minimum of (8.40), eliminates the Goldstone fields.

To investigate the form of the gauge fields when the Goldstone fields are eliminated it is again convenient to consider the Noether current, namely

$$j_\mu^a = \frac{\partial L}{\partial A_\mu^a} = e\bar{\psi}\gamma_\mu\sigma_a\psi + e(\sigma_a\phi, D_\mu\phi) + e^2 A_\mu^b(\phi, \sigma_a\sigma_b\phi), \tag{8.44}$$

which after the SSB becomes

$$j_\mu^a = \partial_\mu e(\sigma_a\mathring{\phi}, \theta) + M_{ab}A_\mu^b + \text{nonlinear terms.} \tag{8.45}$$

Using the current conservation law one then has

$$\square\, e(\sigma_a\mathring{\phi}, \theta) + M_{ab}(\partial_\mu A_\mu^a) = \text{nonlinear terms.} \tag{8.46}$$

It follows that for the asymptotic (free) fields the vanishing of the Goldstone components $(\sigma_a\mathring{\phi}, \theta)$ implies that

$$M_{ab}(\partial_\mu A_\mu^a) = 0. \tag{8.47}$$

This condition is automatically satisfied by the massless gauge fields, but for the *massive* gauge fields it implies that $p_\mu A_\mu = 0$, or that A_0 vanishes in the rest-frame. Thus, just as in the abelian case, there exists a gauge in which both the Goldstone fields and the timelike components of the massive gauge field vanish. This result constitutes the Higgs mechanism for the non-abelian case.

The gauge in which the Goldstone fields and the timelike components of the gauge fields vanish is called the *physical* or *unitary* gauge for the same reasons as in the abelian case. In this gauge the broken Lagrangian takes the somewhat simpler form

$$-L(A_\mu, \theta) = \tfrac{1}{4}F_{\mu\nu}^a F_{\mu\nu}^a + \tfrac{1}{2}(\partial_\mu\theta_i)^2 + A_\mu^r A_\mu^s[\tfrac{1}{2}M_{rs} + g_{rs}^i\theta_i + g_{rs}^{ij}\theta_i\theta_j] + V(\theta_i), \tag{8.48}$$

where M_{rs} is the gauge mass matrix, g_{rs}^i and g_{rs}^{ij} are the quantities $e^2(\sigma_r \sigma_s \overset{\circ}{\phi})_i$ and $e^2(\sigma_r \sigma_s)_{ij}$ respectively, and the index i runs only over the non-Goldstone components. Unfortunately, the physical gauge is one in which computations are difficult, especially in the quantized case, so for many purposes, such as proving renormalizability, or making perturbative expansions, other gauges are used.

8.6   Completeness of the Higgs mechanism

An interesting question that might be asked is whether it is necessary to gauge the whole symmetry group G of a potential for the Higgs mechanism to work. For simple groups the answer is 'yes', which can be seen from the following slightly more general lemma:

LEMMA

Let (G, H) be the symmetry and little algebras of a spontaneously broken potential, and suppose that the Higgs mechanism operates. Then G must be a direct sum of the form $G = G_g + G_h$, where G_g is completely gauged and G_h remains unbroken (i.e. G_h is a spectator algebra).

In other words, all of G is gauged except for a part G_h which takes no part whatsoever in the process (e.g. the Poincaré algebra in a spontaneous breakdown of an internal symmetry).

Proof: If H is the little algebra, then its orthogonal complement $H_\perp$ in G spans the Goldstone directions. Thus the smallest algebra which will contain the Goldstone directions, and hence implement the Higgs mechanism, is the closure K of $H_\perp$ with respect to commutation, i.e. $K = H_\perp, [H_\perp, H_\perp], [H_\perp, [H_\perp, H_\perp]], \ldots$. Now since H is compact it fully reduces G, and thus $H_\perp$ is invariant with respect to H. It then follows from the Jacobi identity that each of the subspaces $H_\perp [H_\perp, H_\perp], \ldots$, in K, and thus K itself, is invariant with respect to H. On the other hand, K is invariant with respect to $H_\perp$ by definition. Thus K is invariant with respect to $G = H + H_\perp$, i.e. K is an invariant subalgebra of G. Since G is compact it follows that G is a direct sum of the form $G = K + K_\perp$. But since K is the smallest algebra that contains the Goldstone directions, K must be gauged for the Higgs mechanism to work, and since $K_\perp$ contains no Goldstone directions $K_\perp$ must be contained in H and thus must remain unbroken.

This lemma will be useful in discussing the gauge-hierarchy problem in section 12.4.

Exercises

8.1. Compute the $SU(n)$ gauge-field masses when the representation for the scalar field is:

    (i)   the fundamental, and $\overset{\circ}{\phi}$ is the vector $\mu(0, 0, 0, ..., 0, 1)$;

    (ii) the adjoint, and $\overset{\circ}{\phi}$ is the matrix μ diag$(1, 1, 1, ..., 1-n)$.

8.2. Assigning the fermions and scalars to the fundamental and adjoint representations of $SU(n)$, write down the Yukawa couplings (and $SU(n)$-invariant mass terms) and compute the fermion mass-spectrum for $\overset{\circ}{\phi}$ as in exercise 8.1.

8.3. Identify the Goldstone and polar fields in the two cases of exercise 8.1, and verify the Higgs mechanism in each case.

9
Gauge theory of the non-gravitational interactions

9.1 Introduction

The fundamental physical interactions known at the present time are the gravitational, electromagnetic, strong nuclear and weak nuclear forces respectively. In this chaper it will be shown how the gauge theory of chapters 7 and 8 are used to describe the latter three interactions, and to introduce a certain degree of unification among them.

9.2 Gauge theory of the strong interactions

The early theories of the strong interactions were local Lagrangian theories, with the baryons and mesons playing the role of local fields. Once the non-local nature of the baryons became apparent (through the EM form-factors for example) these theories became untenable and there came a period in which local Lagrangians were abandoned in favour of more general structures, such as analytic S-matrix theory and current algebra. The advent of quark theory, however, has permitted a return to local Lagrangians with the quarks taking over the role of the fermions and non-abelian gauge fields (gluons) that of the mesons, and this is the theory that is used at present. In this theory the baryons and mesons are regarded as composite fields, with interactions which are themselves non-local but are generated by the local quark–gluon interactions (Isgur and Karl, 1983). An analogy, perhaps, is the case of non-local molecular interactions generated by local quantum electrodynamics.

There are a number of reasons why non-abelian gauge fields are chosen as gluons (so called because they generate the quark binding forces), namely: (i) the success of gauge theory for the electroweak interactions; (ii) the fact that gravitation is also a gauge theory; (iii) the pre-existence of an exact rigid symmetry (colour), ripe for gauging; and (iv) the asymptotic freedom of non-abelian gauge theories. The last reason is actually two-fold: first, the asymptotic freedom itself explains the almost free behaviour of quarks found experimentally in deep inelastic scattering

101

(Buras, 1980). Second, the converse of asymptotic freedom, namely the *increase* of the coupling constant g with decreasing energy (increasing distance) provides a qualitative explanation of quark confinement (Bander, 1981). (Although the continued increase of g for large values of g is an assumption, since it cannot be checked by perturbative methods, it is strongly supported by non-perturbative lattice calculations.) Finally, a list of technical reasons why non-abelian gluons are preferable to abelian ones has been given by Fritzsch, Gell-Mann and Leutwyler (1973).

Although all these arguments are rather qualitative, they are so plausible that they have been almost universally accepted (see for example Fritzsch and Minkowski (1981) and Wilczek (1982)). Once they are accepted, the strong interaction Lagrangian is almost unique, namely,

$$-L(x) = \tfrac{1}{4}\mathrm{tr}\,(F_{\mu\nu}\,F_{\mu\nu}) + \sum_{\alpha} \bar{q}_\alpha \rlap{/}D q_\alpha + \sum_{\alpha,\beta} \bar{q}_\alpha m_{\alpha\beta} q_\beta, \qquad (9.1)$$

where $F_{\mu\nu}$ is the gauge field associated with the colour group (presumably $SU(3)$), the sum is over all quark *flavours*, quark colour indices being suppressed, and $m_{\alpha\beta}$ is the quark-mass matrix. Note that $m_{\alpha\beta}$ is the only quantity in (9.1) which breaks the flavour symmetry (and also the space–time conformal symmetry). No scalar fields are included because (*a*) the colour symmetry is exact and need not be spontaneously broken, and (*b*) scalar fields diminish asymptotic freedom.

The now-conventional strong-interaction Lagrangian (9.1) is very similar to the quantum electrodynamics (QED) Lagrangian, and for this reason the strong interaction theory (9.1) is often called quantum chromodynamics (QCD). In fact the only formal difference between QCD and QED is that $F_{\mu\nu}$ contains the nonlinear term $[A_\mu, A_\nu]$ and that there are r gluons (gauge fields A_μ) instead of one photon, where r is the order of the group ($r = 8$ for $SU(3)$). However, this formal difference produces important physical differences, such as asymptotic freedom, and, if the reasons for adopting QCD are correct, it should produce the most important difference of all, namely a mechanism for quark confinement (Becher, Bohm and Joos, 1984; Wilczek, 1982).

The analogy with QED means that many of the results of QCD which depend only on perturbation theory (modulo some background potential which is put in to allow for confinement) can be obtained from the analogous results for QED, and, in general, the perturbative results for QCD are in good agreement with experiment (Mueller, 1981; Reya, 1981). Of course, it would be desirable to have direct, non-perturbative evidence for the correctness of QCD, but, on account of confinement and the nonlinearity of the theory, it is not easy to obtain such evidence. One direct prediction of QCD is the existence of glueballs, i.e. colourless multi-gluon bound states, but the unambiguous identification of such states is difficult (Fishbane and Meshkov, 1985). Another prediction is that the momentum plots for hadron scattering should exhibit long narrow bulges, called jets, due to the formation inside the hadrons of quark–quark and quark–gluon resonances. Jets which can be interpreted in this manner have been clearly seen, and they are generally regarded as good evidence for the correctness of the theory (Söding and Wolf, 1981; Bassetto, Ciafolini and Marchesini, 1983). The desire to check QCD directly, and the fact that this cannot be done perturbatively, has led to a rapid growth in lattice gauge theory, where non-perturbative, computer-based, computations can be made (Creutz, 1983; Kogut, 1983; Rebbi, 1983). The results of those lattice calculations which have been made to date support QCD.

9.3   Standard model of the weak and electromagnetic interactions

The history of the weak interactions is quite different to that of the strong. In fact, once the parity violation and the chiral, or vector-axial vector (V-A) structure of the weak Noether currents became apparent (Marshak, Riazud-din and Ryan, 1969), the idea that the weak interactions were mediated by an intermediate vector meson $W^{\pm}_{\mu}$ soon followed. However, in contrast to the photon, the $W^{\pm}_{\mu}$-meson had to be charged, parity violating and massive, and it was not known how to construct a self-consistent renormalizable theory for such fields. Thus for the weak interactions the problem was not so much to consider the idea of using vector mesons as to implement this idea.

This problem was solved in three stages. First, the Yang–Mills gauge theory provided a natural method of introducing charges for the vector mesons. Next, the discovery of spontaneous symmetry breaking provided a mechanism for introducing the mass. Finally, the technical problem of proving renormalizability was solved by using dimensional regularization (Leibrandt, 1975) and functional integration. The proofs of renormalization may be found in the literature on unified gauge theory in the suggestions for further reading at the end of this book.

Because of the discrepancy between the vectorial (V) structure of electromagnetism and the chiral (V-A) structure of the weak interactions, the minimal gauge group necessary to incorporate the photon and charged W-meson without introducing new fermions is not $SU(2)$ but $U(2)$, and a model based on this group is now the standard one. It was first proposed (without SSB) by Glashow (1961) and Salam and Ward (1964), and the SSB was introduced into it by Weinberg (1967) and Salam (1968). Since $U(2)$ has four generators the model contains a massive neutral meson Z_μ^0 as well as $W_\mu^\pm$, and the model first came to be accepted when the corresponding neutral current J_μ^0 was found experimentally, and the structures of the currents J_μ^0 and $J_\mu^\pm$ were shown to be in excellent agreement with experiment (six experimental numbers described by one new parameter). For reviews see Hung and Sakurai (1981), Kim *et al.* (1981), Bilenky and Hösek (1982)). The recent observations of the Z_μ^0 and $W_\mu^\pm$ themselves (Arnison *et al.* 1983; Bagnera and Banner *et al.* 1983) have removed any lingering doubts.

In terms of the gauge fields $\mathbf{A}_\mu$ and A_μ for the $SU(2)$ and $U(1)$ subgroups of $U(2)$, the photon and vector mesons $W_\mu^\pm$, Z_μ^0 may be written as

$$W_\mu^\pm = \frac{1}{\sqrt{2}}(A_\mu^{(1)} \pm iA_\mu^{(2)}), \qquad \begin{aligned} A_\mu^{\text{em}} &= A_\mu^{(3)} \cos\theta + A_\mu \sin\theta, \\ Z_\mu^0 &= A_\mu \cos\theta - A_\mu^{(3)} \sin\theta, \end{aligned} \tag{9.2}$$

where the 'weak angle' θ is the new parameter just mentioned. Similarly the electric charge operator Q may be written as a linear combination of the generators σ_3 of $SU(2)$ and y of $U(1)$, and it is conventional to normalize y so that

$$Q = \sigma_3 + \tfrac{1}{2}y, \quad (2\text{tr}\sigma_3^2 = 1). \tag{9.3}$$

With this normalization y is integer for the leptons and unity for the universal scalar field (below). Since σ_3 and y are generators, (9.3) holds for all representations of $U(2)$.

The fermions are assigned to representations of $U(2)$ by assembling them in the pairs (a, b) which are coupled by $W_\mu^\pm$ (in practice this is $(a, b) = (\nu^e, e)\,(\nu^\mu, \mu)\,(\nu^\tau, \tau)$ for the leptons and, with a modification to be discussed later, $(a, b) = (u, d)\,(c, s)\,(t, b)$ for the quarks). Then since the coupling to $W_\mu^\pm$ is (left-handedly) chiral, the left- and right-handed components of (a, b) are assigned to an $SU(2)$ doublet and two $SU(2)$ singlets respectively. Symbolically $(a, b) = \{(a_L, b_L)\,a_R, b_R\}$ where, for massless neutrinos the right-handed components are zero. The $U(1)$ assignment is a little trickier. Since the electric charge is parity-invariant and $\Delta Q = \pm 1$ for the $W_\mu^\pm$ interaction, one sees that the charges of (a, b) are fixed up to an overall charge, i.e. $Q = \tfrac{1}{2}\{(q+1, q-1)\,q+1, q-1\}$. The overall charge q is determined from the electric charge of one of the members of the multiplet and turns out to be $q = -1$ for leptons and $q = \tfrac{1}{3}$ for quarks.

Table 4. *The $U(2)$ assignments of the first generation $\{(\nu^e, e)-(u, d)\}$ of quarks and leptons. The assignments for the other two generations $\{(\nu^\mu, \mu)-(c, s)\}$ and $\{(\nu^\tau, \tau)-(t, b)\}$ are similar.*

	(ν_L^e, e_L)	$\nu_R^e(?)$	e_R	$(u_L d_L)$	u_R	d_R
σ_3	$(\ \tfrac{1}{2}, -\tfrac{1}{2})$	0	0	$(\tfrac{1}{2}, -\tfrac{1}{2})$	0	0
y	$(-1, -1)$	0	-2	$(\tfrac{1}{3}, \ \tfrac{1}{3})$	$\tfrac{4}{3}$	$-\tfrac{2}{3}$
Q	$(\ 0, -1)$	0	-1	$(\tfrac{2}{3} - \tfrac{1}{3})$	$\tfrac{2}{3}$	$-\tfrac{1}{3}$

From (9.3) it then follows that the $U(1)$ charge assignment is $y = \{(q, q), q+1, q-1\}$ where $q = -1$ for leptons and $q = \tfrac{1}{3}$ for quarks.

There is only one scalar field in the $U(2)$ model. This is a $U(2)$ doublet, and, so that it can have a Yukawa interaction (which couples left to right and therefore has $\Delta y = \pm 1$) the doublet must have $y = 1$. Thus its $U(1)$ charge is unity.

Using the first pair of leptons and quarks, (ν^e, e) and (u, d), as examples, the above assignments may be tabulated explicitly as in table 4. Note that y is $SU(2)$ and Q is parity-invariant.

Before going on to the $U(2)$ Lagrangian there are two comments which may be made concerning the normalization of y. First, although it has become traditional to use y, a more natural quantum number would be $m = 3y$ since this would be integer for all fields and would correspond to the character of the $U(1)$ representation. In particular, using m, one sees from table 4 that $m = 0, 1$ (mod 2) for the $SU(2)$ singlets and doublets respectively, and from section 5.5 one recalls that this is just the condition that the assignment be $U(2)$ rather than $SU(2) \times U(1)$. Second, if $U(2)$ is embedded in a simple group with y as generator, then in general y will not be normalized in the same way as σ_3 (tr $y^2 \neq 2$). In that case it is natural to rescale the operator y to an operator σ which is so normalized, and the charge formula (9.3) becomes

$$Q = \sigma_3 + c\sigma \qquad (9.4)$$

where $2\mathrm{tr}\ \sigma_3^2 = 2\mathrm{tr}\ \sigma^2 = 1$, and c is a constant which is uniquely determined by the simple group. Note that (9.4) implies that

$$\mathrm{tr}\ Q^2 = (1+c^2)\ \mathrm{tr}\ Q_3^2, \qquad (9.5)$$

which relates c^2 to the sum of the squares of the electric and $SU(2)$ charges in *any* representation of the simple group. In particular, if for any group (such as the $SU(5)$ or $SO(10)$ groups of grand unification) the fermions

listed in the above table (with or without ν_R) form a complete representation, then (9.5) applies to these fermions, and one sees that $c^2 = \frac{5}{3}$.

After this digression, let us consider the $U(2)$ Lagrangian. Once the
scalar and fermion assignments have been made it is determined up to the
value of a few parameters. The part involving the gauge field takes the form

$$-L_{\text{gauge}} = \tfrac{1}{4}\mathbf{F}_{\mu\nu}\cdot\mathbf{F}^{\mu\nu} + \tfrac{1}{4}F_{\mu\nu}F^{\mu\nu} + \tfrac{1}{2}(D_{\mu}\phi)^2 + \sum_{\alpha}\bar{\psi}\mathcal{D}\psi, \qquad (9.6)$$

where the sum is over all fermion multiplets and the covariant derivative
is

$$D_{\mu} = \partial_{\mu} + g\mathbf{A}_{\mu}\cdot\boldsymbol{\sigma} + \tfrac{1}{2}fA_{\mu}y. \qquad (9.7)$$

The matter–matter part of the Lagrangian takes the form

$$-L_{\text{matter}} = \sum_{\alpha}G_{\alpha}(\bar{\psi}_{\alpha L}, \phi)\,\psi_{\alpha R} + \text{herm. conj.} + \lambda((\phi,\phi)-c^2)^2, \qquad (9.8)$$

where there is a separate (complex) Yukawa coupling constant G_{α} for each
fermion field, the inner product is with respect to $SU(2)$, and the last term
(which involves only two parameters, λ and c) is the scalar potential. Note
that there are no $SU(2)$-invariant mass terms, so all the fermion masses
must be generated by the SSB as described in chapter 8.

In order to determine the physical fields A_{μ}^{em} and Z_{μ}^{0} (i.e. to relate the
weak angle θ to the constants f and g) and to display their currents, one
replaces the operator y in the covariant derivative by $2(Q-\sigma_3)$ to obtain

$$D_{\mu} = \partial_{\mu} + \frac{g}{\sqrt{2}}W_{\mu}^{\pm}\sigma^{\mp} + (gA_{\mu}^{(3)} - fA_{\mu})\sigma_3 + fA_{\mu}Q. \qquad (9.9)$$

The fields A_{μ}^{em} and Z_{μ}^{0} are then identified by noting that, since A_{μ}^{em}
preserves parity, it cannot couple to σ_3. Hence the coefficient of σ_3 in (9.9)
must be the orthogonal complement of A_{μ}^{em}, namely Z_{μ}^{0}. Changing to A_{μ}^{em}
and Z_{μ}^{0} by using this identification (i.e. $\tan\theta = f/g$) in (9.2) one finds that

$$D_{\mu} = \partial_{\mu} + \frac{g}{\sqrt{2}}W_{\mu}^{\pm}\sigma^{\mp} + (g^2+f^2)^{\frac{1}{2}}Z_{\mu}^{0}\left(\sigma_3 - \frac{f^2}{f^2+g^2}Q\right) + \frac{fg}{(g^2+f^2)^{\frac{1}{2}}}A_{\mu}^{em}Q.$$
$$(9.10)$$

From (9.10) (and (9.2)) one then has the identifications

$$\tan\theta = \frac{f}{g}, \quad e_0 = \frac{fg}{(f^2+g^2)^{\frac{1}{2}}} \quad \Leftrightarrow \quad e_0 = f\cos\theta = g\sin\theta, \qquad (9.11)$$

where e_0 is the usual EM coupling constant. Thus for the (ν^e, e)-system, for example, the gauge-fermion interaction is

$$\frac{g}{\sqrt{2}}(\bar{\nu}\gamma_\mu \bar{e})\, W_\mu^+ + \text{herm. conj.} + e_0(\bar{e}\gamma_\mu e)\, A_\mu^{em}$$

$$+ \tfrac{1}{4}(g^2+f^2)^{\frac{1}{2}}\left\{2\bar{\nu}\gamma_\mu \nu + \bar{e}\gamma_\mu \gamma_5 e + \frac{3f^2-g^2}{f^2+g^2}\bar{e}\gamma_\mu e\right\} Z_\mu^0, \quad (9.12)$$

where the first two terms are the traditional weak and electromagnetic couplings (Marshak, Riazud-din and Ryan, 1969) and the last term is the new coupling induced by $U(2)$.

The masses of the W_μ and Z_μ^0 fields after SSB are easily computed as a special case of the formula (8.8), and are

$$M_W = g|\mathring{\phi}|, \quad M_Z = (g^2+f^2)^{\frac{1}{2}}|\mathring{\phi}|. \quad (9.13)$$

The value of $|\mathring{\phi}|$ may be obtained independently from β-decay, in which $|\mathring{\phi}|^2 = (2G)^{-1}$ where G is the traditional 4-fermi coupling constant, and the values of g and f may be obtained independently from e_0 and the experiments on the neutral current of (9.12), mentioned earlier. Substituting these values in (9.13) one obtains $M_W \approx 80$ GeV and $M_Z \approx 92$ GeV which is in excellent agreement with the experimental values: $M_W \approx 81 \pm 3$ and $M_Z \approx 94 \pm 6$ GeV.

9.4 $S(U(3) \times U(2))$ theory of the non-gravitational interactions: generations

On combining the results of the last two sections one sees that all the known non-gravitational interactions can be described by a gauge theory based on independent $SU(3)$ and $U(2)$ groups. The matter fields consist of the fields listed in table 4, where the quarks are $SU(3)$ triplets and the leptons and the universal scalar are $SU(3)$ singlets. From table 4 one sees that all the matter fields satisfy the relations

$$t = 3y \,(\text{mod } 3), \quad 2\sigma_3 = 3y \,(\text{mod } 2), \quad (9.14)$$

and from section 5.3 this shows that the combined group is $S(U(3) \times U(2))$ rather than $SU(3) \times U(2)$. Similarly, since $Q = 3y \,(\text{mod } 3)$, the unbroken gauge symmetry is actually $U(3)_{cem} = SU(3)_c \times U(1)_{em}/Z_3$, rather than the more familiar $SU(3)_c \times U(1)_{em}$. Thus the established gauge group of the non-gravitational interactions is $S(U(3) \times U(2))$, where $U(2)$ describes the electroweak interactions and $U(3)$ the electrostrong ones.

There are, however, two points which should be observed. First, the $U(1)$ subgroup of $S(U(2) \times U(1))$ does not automatically satisfy the renormalizability condition due to the ABJ anomaly discussed in section 7.4, since from (7.26) the $U(1)$ condition is that the sum of all $U(1)$ charges (or equivalently from (9.3) the sum of all electric charges) be zero. For the lepton and quark pairs the sums of the electric charges are -1 and $\frac{1}{3}$ respectively, and since the quarks are colour-triplets, one then sees that the ABJ condition requires that the number of lepton and quark pairs should be equal. This is a rather remarkable result because it establishes for the first time a direct relationship between the baryons and leptons, and it has important implications for grand unification and for cosmology (see suggestions for further reading at the end of this book), where it could, for example, explain the apparent electron–proton symmetry of the universe. The lepton–quark relation is usually assumed to be pairwise, i.e. that the leptons and quarks occur naturally in sets such as $\{(\nu^e, e) \, (u^a, d^a)\}$ $\{(\nu^\mu, \mu) \, (c^a, s^a)\}$ and $\{(\nu^\tau, \tau), (t^a, b^a)\}$, each of which is anomaly free. A renormalizable $S(U(3) \times U(2))$ theory can be built with each set separately, and for this reason such sets are known as $S(U(3) \times U(2))$ *generations*.

The second point is that, as they stand, the assignments of the quarks forbid inter-generation decays, in contradiction with experimental evidence (for example the fact that $\sin \theta_C \neq 0$ in the $SU(3, f)$ weak current of section 6.3). More generally, the experimental evidence is that the *charged* currents connect generations and the *neutral* currents do not, and an elegant way to incorporate this situation into the $U(2)$ theory is to generalize the $SU(2)$ generators as follows:

$$\sigma_3 \to \mathbb{1} \times \sigma_3, \quad \sigma_+ \to U \times \sigma_+, \quad \sigma_- \to U^\dagger \times \sigma_-, \tag{9.15}$$

where U is a $U(2)$-invariant inter-generation unitary matrix ($n \times n$ for n generations). The generalization (9.15) is called the GIM mechanism (Glashow, Iliopolous and Miani, 1970) and the matrix U the KM-matrix (Kobayashi and Maskawa, 1973). For two generations the KM-matrix is 2×2 and is nothing but a rotation through the Cabibbo angle θ_C. In fact, historically, it was just the requirement that the $SU(3, f)$ Cabibbo angle in section 6.3 be replaced by a rotation matrix that forced the generalization of $SU(3, f)$ to $SU(4, f)$ and thus introduced charm. For more than two generations the $\frac{1}{2}n(n+1)$ independent phases for the matrix elements exceed the $2n-1$ independent phases for the base elements on either side by $\frac{1}{2}(n-1)(n-2)$ and hence the KM-matrix may have $\frac{1}{2}(n-1)(n-2)$ intrinsically complex entries. This fact enables the KM-matrix to incorporate CP (time-reversal) violations for $n \geqslant 3$. For a review of the KM-matrix and quark mixing see Chau (1983).

The existence of the KM-matrix shows that the 'strong' quarks (the ones for which the strong-interaction mass matrix (8.1) is diagonal) are not the same as the 'weak' quarks (those assigned to $U(2)$), but are unitary transforms of them. Since both sets have the same electric and $U(1)$ charges the unitary matrix V connecting them must be a direct sum $V = V^+ + V^-$ where the $\pm$ denotes the σ_3 value. Thus

$$\begin{pmatrix} u \\ c \\ t \end{pmatrix}_{\text{strong}} = (V^+) \begin{pmatrix} u \\ c \\ t \end{pmatrix}_{\text{weak}}, \quad \begin{pmatrix} d \\ s \\ b \end{pmatrix}_{\text{strong}} = (V^-) \begin{pmatrix} d \\ s \\ b \end{pmatrix}_{\text{weak}}$$

and it is easy to verify that the KM-matrix is just the product V^+V^-. There is no corresponding change of basis for the leptons since they have no strong interaction, and because the neutrinos are massless.

Taking the above two points into consideration, the $S(U(3) \times U(2))$ assignment of table 4 is correct, and it completely defines the $S(U(3) \times U(2))$ model (up to the values of the parameters). In principle, this model should describe *all* non-gravitational phenomena, but, of course, it is very difficult to extract the experimental information from the strong part. Furthermore, the existance of so many free parameters, notably in the fermion mass matrix and the KM-matrix, shows that the model is not complete.

Exercises

9.1. Using standard Feynman graph analysis, show that in the one-loop approximation the contribution to the fermion and scalar self-energy and the vacuum polarization due to the fermion and scalar loops is independent of the structure constants (and is thus the same for QCD and r copies of QED), but that, in contrast to QED, QCD has a gauge-field loop contribution to gauge-field self-energy and the vacuum polarization.

9.2. Construct an $S(U(2)_L \times U(2)_R)$ model with left- and right-handed quarks as $U(2)_L$ and $U(2)_R$ doublets, and assign scalars so that it breaks down to the standard model

9.3. Using a Yukawa interaction of the form (9.8) compute the masses of the first generation fields (ν, e, u, d) in terms of the Yukawa couplings and the vacuum value ϕ of the scalar field.

9.4. Show that by a suitable choice of basis any unitary 3×3 matrix, and hence the 3×3 KM-matrix, may be written as

$$\begin{bmatrix} \cos\alpha & -\sin\alpha & 0 \\ \sin\alpha & \cos\alpha & 0 \\ 0 & 0 & 1 \end{bmatrix} \begin{bmatrix} 1 & 0 & 0 \\ 0 & e^{i\epsilon} & 0 \\ 0 & 0 & 1 \end{bmatrix} \begin{bmatrix} \cos\beta & 0 & -\sin\beta \\ 0 & 1 & 0 \\ \sin\beta & 0 & \cos\beta \end{bmatrix} \begin{bmatrix} 1 & 0 & 0 \\ 0 & \cos\gamma & -\sin\gamma \\ 0 & \sin\gamma & \cos\gamma \end{bmatrix}$$

where $\alpha, \beta, \gamma, \epsilon$ are real and thus $e^{i\epsilon}$ is the only complex entry.

10
Grand unification

10.1 Introduction

The $S(U(3) \times U(2))$ gauge theory of the non-gravitational interactions is not fully unified, in the sense that it contains three independent gauge couplings, and that the leptons and quarks, and more generally the different generations, belong to independent representations, and this suggests that $S(U(3) \times U(2))$ is only a subgroup of some larger, preferably simple, gauge group G. Gauge theories based on such larger groups G, which unify the $SU(3)$, $SU(2)$ and $U(1)$ subgroups of $S(U(3) \times U(2))$ are called grand unified theories, or GUTs. Such theories were first considered by Pati and Salam (1973) and Georgi and Glashow (1974) and will be the subject of this chapter. The novel feature of GUTs is that they involve proton decay.

10.2 Renormalization group preliminary

Before GUTs can even be considered, there is an important question that must be answered, namely, how can there be a single overall gauge coupling constant in a single group G when the present values of the three coupling constants in $S(U(3) \times U(2))$ differ by so much – certainly more than could be reasonably accounted for by group factors such as Clebsch–Gordan (CG) coefficients.

As first discussed by Georgi, Quinn and Weinberg (1974) the answer to this question lies in the renormalization group equations of section 7.5. To see this, consider the integrated one-loop equation (7.29),

$$f(\mu) = f(m) - 2b \ln (\mu/m), \quad f(\mu) = g(\mu)^{-1}, \tag{10.1}$$

where μ is the scale parameter, b is the leading coefficient in the β-coupling (to fields of mass m, say) and $f(m)$ is the experimental value of $f(\mu)$ for $\mu = m$. Now suppose that heavier fields (of mass M and b-constant B, say) which had been neglected in (10.1) were to be taken into account. Then

110

$f(m)$ would not change since it is an experimental value, but (10.1) would change to

$$F(\mu) = f(m) - 2b \ln(\mu/m) - 2B \ln(\mu/M), \quad F(\mu) = G^{-1}(\mu), \quad (10.2)$$

where $G(\mu)$ is the new effective coupling constant, and the last term is the one-loop contribution of the fields M. Thus $F(\mu)$ would satisfy the modified renormalization group equation

$$\mu \frac{\mathrm{d}}{\mathrm{d}\mu} F = -2(b + B), \qquad (10.3)$$

where $F(M) = f(M)$, (and for higher loops, this equation with $(b + B)$ replaced by $\beta(b, B)$). This is a well-known result (Symanzik, 1969; Appelquist & Carrazone, 1975; Weisberger, 1981) and is normally used to show that higher-mass fields can be neglected at low energies, because, although the terms $\ln m/M$ are not negligible like the powers of (m/M), they can be absorbed in the renormalization of the coupling constants. For this reason the result is sometimes referred to as the 'decoupling theorem'.

Consider now the special case of $S(U(3) \times U(2))$ and let $g_i = e_i^2$ where e_i, $i = 1, 2, 3$ are the gauge coupling constants for the $U(1)$, $SU(2)$ and $SU(3)$ subgroups. Their renormalization group equations (10.1) are of the form

$$f_i(\mu) = f_i(m) - 2b_i \ln(\mu/m), \quad f_i = g_i^{-1} = e_i^{-2}, \qquad (10.4)$$

where, because the ln-dependence is not strong, all the $S(U(3) \times U(2))$ masses m are assumed to be the same, $m \approx 10$ GeV, say. Now suppose there is a single GUT group G which breaks down to $S(U(3) \times U(2))$ at some larger mass, M say $(|\overset{\circ}{\phi}| \sim M)$. Then the analogue of (10.2) is

$$F_i(\mu) = f_i(\mu) - 2B_i \ln(\mu/M). \qquad (10.5)$$

On the other hand, G-invariance requires that there be only one overall gauge coupling, G say, so $F_i(\mu) = F(\mu)$, and that there be only one overall β-function, so $b_i + B_i = B$. For example, for pure gauge theory b_i are the Casimirs for $U(1)$, $SU(2)$ and $SU(3)$ and B is the Casimir for G. It then follows from (10.5) that the combinations $f_i - 2b_i \ln(m/M)$ are independent of i,

$$f_i - 2b_i \ln(\mu/m) = F - 2B \ln(M/\mu), \quad i = 1, 2, 3. \qquad (10.6)$$

Thus, although the G-couplings $F_i = F$ are the same, the $S(U(3) \times U(2))$ couplings f_i may differ by logarithmic amounts.

The problem is to see whether the observed differences for $SU(3)$, $SU(2)$ and $U(1)$ can be explained in this way and to estimate the scale of M (which must be very large on account of the slow variation of the logarithm). For this purpose it is convenient to eliminate $F(\mu)$ from (10.6) and write the two remaining equations in the form

$$f_1(b_2-b_3)+f_2(b_3-b_1)+f_3(b_1-b_2) = 0, \tag{10.7}$$

$$\ln(M/m)^2 = (f_i-f_j)(b_i-b_j)^{-1} \quad (\text{any } i \neq j). \tag{10.8}$$

Then equation (10.7) is a compatibility equation for the low-energy parameters alone, and (10.8) determines M. Note that (10.7) and (10.8) depend only on the differences (b_i-b_j). This is fortunate because then the fermion-loop contributions drop out completely (see exercise 9.1), and from (7.31)

$$(b_2-b_3) = (\tfrac{11}{3}+\tfrac{1}{6}s)\,K, \quad (b_1-b_3) = (11+\tfrac{1}{10}s)\,K, \tag{10.9}$$

where $K = (8\pi)^{-2}$. Furthermore, unless the number of scalar fields is very large, the gauge-field contributions dominate and one may write

$$\tfrac{1}{3}(b_1-b_3) = \tfrac{1}{2}(b_1-b_2) = b_2-b_3 = \tfrac{11}{3}K. \tag{10.10}$$

Thus the consistency condition (10.7) may be written as

$$f_1-3f_2+2f_3 = 0, \quad \text{or} \quad 3(g_1/g_2)^2 = 1+2(g_1/g_3)^2. \tag{10.11}$$

The ratio $(g_1/g_2)^2$ may be obtained by using the identification

$$\ldots g_1\,\sigma+g_2\,\sigma_3\ldots(\text{for } G) = \ldots e_0\left(\frac{c\sigma}{\cos\theta}+\frac{\sigma_3}{\sin\theta}\right)\ldots(\text{for } U(2)), \tag{10.12}$$

which follows from (9.4) and (9.11), where e_0 is the EM coupling constant and θ is the weak angle of section 9.2. The ratio $(g_1/g_3)^2$, although not known exactly, is small (say of order 10^{-1}). Thus the consistency condition is

$$3c^2\tan^2\theta \gtrsim 1. \tag{10.13}$$

Since $\tan^2\theta \approx \tfrac{1}{4}$, one has $c^2 \gtrsim \tfrac{4}{3}$ which is not far from the value $c^2 = \tfrac{5}{3}$ obtained from (9.5) using the known fermions. Thus the consistency condition is roughly satisfied by the known values of the low-energy coupling constants. Note that in the exact symmetry limit $g_1 = g_2 = g_3$ and hence $c^2\tan^2\theta = 1$. Thus the renormalization group makes a considerable difference.

To estimate the scale of M it is convenient to use (10.8) for $i,j = 1, 2$.

$$\ln\left(\frac{M}{m}\right) = \frac{(f_1-f_2)}{2(b_1-b_2)} = \frac{6\pi}{11}\left(\frac{4\pi}{e_0^2}\right)\left(\frac{\cos^2\theta}{c^2}-\sin^2\theta\right), \tag{10.14}$$

which for $\pi \approx \tfrac{22}{7}$, and $\sin^2\theta \approx \tfrac{2}{9}$ yields $\ln(M/m) \approx 26(7-2c^2)/2c^2$. This shows that M is fairly sensitive to the value of c^2, and for $c^2 \approx \frac{5}{3}$ is of order $M \approx m \exp 28 \approx m10^{14} \approx 10^{15}$ GeV. This is an enormous energy by present-day scales, but it does not seem so outlandish when it is noted that it is less than the Planck mass (10^{19} GeV) at which gravitation becomes important, and that it is of the order required to keep the proton stable (next section). In fact, there is a small 'window' ($\approx 10^{15}-10^{19}$ GeV) high up on the energy scale which is bounded below by the present experimental information on proton decay and above by the Planck mass, and most estimates of the grand unification scale based on the renormalization group land in or about this window. Because of this, the scale $M \approx 10^{15}$ GeV has come to be accepted as the energy scale (value of $|\overset{\circ}{\phi}|$) at which grand unification breaks down to $S(U(3) \times U(2))$. (Buras *et al.*, 1978; Langacker, 1981).

10.3 Minimal GUT ($G = SU(5)$)

Once the problem of the different coupling constants is out of the way, grand unification may be considered in more detail, and the primary problem, of course, is to decide on the GUT group G. Before going on to this choice, and the general problems of GUTs, it is convenient to first fix the ideas by means of a specific example. The simplest example is $G = SU(5)$, because $SU(5)$ is the smallest simple group which contains $S(U(3) \times U(2))$, and even though the $SU(5)$ model appears to be ruled out by present experiments (at least as an ultimate GUT) it is still of pedagogical value.

The gauge fields for the $SU(5)$ model may be written as a 5×5 traceless hermitian matrix

$$A_\mu = A_\mu \sigma + \left[\begin{array}{c|c} A_\mu^{ab} & X_\mu^{ai} \\ \hline \overline{X}_\mu^{ia} & W_\mu^{ij} \end{array} \right], \qquad (10.15)$$

where $\sigma = (\frac{1}{15}) \operatorname{diag}(-2-2-2, +3, +3)$; $a, b = 1, 2, 3$; $i, j = 1, 2$; A_μ^{ab} are the $SU(3, c)$ gauge fields; and W_μ^{ij} are the electroweak $SU(2)$ gauge fields ($W_\mu^\pm$ and $A_\mu^{(3)}$). Thus there are twelve new gauge fields, X_μ^{ai} and their complex conjugates, and since a is a colour-index they are coloured.

One of the most attractive features of the $SU(5)$ model is that each present generation of fermions fits neatly into its fundamental representations, namely, the 5, 10, 10*, 5*, the fifteen members of each generation belonging to the 5* + 10 and their antifields to the 5 + 10*. To see this more explicitly consider the lowest-lying generation $\{(v^e, e), (u^a, d^a)\}$. The $S(U(3) \times U(2))$ decomposition of the 5* and 10 is

$$5^* = (3^*, 1) + (1, 2^*), \quad 10 = (3^*, 1) + (3, 2) + (1, 1),$$

or, in tensor notation,

$$\psi_\alpha^* = \psi_a^* + \psi_i^*, \qquad \psi_{\alpha\beta} = \underset{\vee}{\psi_{bc}} + \underset{\vee}{\psi_{ai}} + \underset{\vee}{\psi_{ij}}, \qquad a, b, c = 1, 2, 3, \quad i, j = 1, 2,$$

$$(10.16)$$

and from this and the charge formula (9.3) it is easy to see that the assignment must be

$$5^* = (\bar{d}, \nu, e)_L, \quad 10 = (\bar{u}, u, d, \bar{e})_L, \tag{10.17}$$

where $(\bar{d})_L = \overline{(d_R)}$, and the charge conjugate, right-handed, fields are assigned to the $5 + 10^*$.

The scalar field assignments in $SU(5)$ are by no means as trivial as in $S(U(3) \times U(2))$, because not only must the scalars belong to complete $SU(5)$ representations but they must:

(i) break $SU(5)$ to $S(U(3) \times U(2))$ *without* generating fermion masses;
(ii) break $S(U(3) \times U(2))$ to $U(3)$ and thereby generate fermion masses in the usual $U(2)$ manner.

This is because the mass scale of the fermions is of the same order as the mass scale of the second breakdown. To see how (i) and (ii) can be arranged, one recalls from section 8.3 that the Yukawa couplings take the form $(f \times f + f^* \times f^*)$ where f is the fermion representation. Thus for $SU(5)$ the Yukawa couplings are through

$$(5^* + 10) \times (5^* + 10) = (10^* + 15^*) + 2(5 + 45) + (5^* + 50^* + 45^*) \quad (10.18)$$

and its hermitian conjugate. Hence condition (i) will be satisfied if the scalar representation for the first breakdown is *not* included in the (Yukawa) set of representations (10.18) and condition (ii) will be satisfied if the scalar representation for the second breakdown *is* included in (10.18) and at the same time contains the usual $S(U(3) \times U(2))$ scalar, namely an $SU(3)$ singlet, $SU(2)$ doublet with $y = 1$. It turns out that the 24-dimensional adjoint representation is the smallest representation which does not belong to (10.18) and breaks $SU(5)$ to $S(U(3) \times U(2))$ (chapter 12) and the only members of (10.18) with the correct $S(U(3) \times U(2))$ content are

$$5 = \{(\mathbf{1,2})(1) + (3,1)(-\tfrac{2}{3})\}, \quad 45 = \{[(\mathbf{1,2}) + (8,2)](1)$$
$$+ [(3,1) + (3,3) + (\bar{6},1)](-\tfrac{2}{3}) + (\bar{3},1)(\tfrac{8}{3}) + (\bar{3},2)(-\tfrac{7}{3})\}, \quad (10.19)$$

where (a, b) denotes the dimensions of the $SU(3)$ and $SU(2)$ representations respectively, the bracket (y) with one entry denotes the $U(1)$ charge, and the required electroweak submultiplet is in bold.

Thus the full assignment of the scalar fields is to $24+(5+45)$ where the 24 is responsible for $SU(5) \to S(U(3) \times U(2))$ and $(5+45)$ for $S(U(3) \times U(2)) \to U(3)$. The 45 is sometimes omitted for simplicity, but it should be noted that in its absence the masses generated for the electron and the d-quark are the same (see exercise 10.2), which is in disagreement with experiment (even allowing for radiative corrections). Similarly for the other generations.

Once the fermion and scalar assignments have been made the $SU(5)$ Lagrangian is unique up to the values of the parameters. In particular the gauge-fermion interaction takes the form

$$e[\bar{\psi}_\alpha \gamma_\mu \psi_\beta + 2 \overline{\varPsi}_{\alpha\gamma} \gamma_\mu \varPsi_{\beta\gamma}] A_\mu^{\alpha\beta}, \qquad (10.20)$$

and on making a decomposition with respect to quarks and leptons this becomes

$$\sqrt{2}\, e(X_1, X_2) \cdot \left\{ d_R \begin{pmatrix} \nu \\ \bar{e} \end{pmatrix}_L + e_R \begin{pmatrix} u \\ d \end{pmatrix}_L + (\bar{u})_R \times \begin{pmatrix} -\bar{d} \\ \bar{u} \end{pmatrix}_L \right\} + \text{h.c.}$$

$$+ eA^{ab} \bar{q} \sigma_{ab} q + e\mathbf{W} \cdot (\bar{q}\sigma q + \bar{l}\sigma l] - \sqrt{\tfrac{3}{5}}\, eAJ, \quad (10.21)$$

where the dot (inner) and cross products are with respect to colour indices, the space–time indices have been suppressed, and

$$J = \bar{l}l + 2\bar{e}_R e_R + \tfrac{1}{3}\bar{u}_L u_L - \tfrac{4}{3}\bar{u}_R u_R + \tfrac{1}{3}\bar{d}_L d_L - \tfrac{2}{3}\bar{d}_R d_R. \qquad (10.22)$$

The first line in (10.21) describes the interactions which occur in $SU(5)$ but not in $S(U(3) \times U(2))$ and the second line describes the old $S(U(3) \times U(2))$ interactions, but with the three coupling constants which were formerly independent now correlated $(g_3 = g_2 = \sqrt{\tfrac{5}{3}}g_1)$. Note that the constant c of (9.4) is then $\sqrt{\tfrac{3}{5}}$ which agrees with the value of c^2 obtained in section 9.3 using the fermion assignment only. Note also that the X-terms do not conserve quark and lepton number separately, and it is this property that leads to the proton decay mentioned earlier. The gauge-field scalar interaction is constructed with the same covariant derivatives as (10.20), the Yukawa interaction takes the form

$$g\bar{\psi}(10)\,\psi(10)\,s(5) + h\bar{\psi}(10^*)\,\psi(5)\,s(5)$$

$$+ k\bar{\psi}(10^*)\,\psi(5)\,s(45) + \text{herm. conj.}, \quad (10.23)$$

and the scalar potential will be discussed in detail in chapter 12.

Although the $SU(5)$ model does not conserve quark and lepton number separately, it preserves the combination $(B-L)$ where $B = $ baryon number $= 3 \times$ quark number and $L = $ lepton number, provided an appropriate value of $(B-L)$ is assigned to the boson fields. This can be seen as follows: let τ be the Noether charge for the $U(1)$ group which corresponds to (rigid) phase changes of each of the complex fields $\psi(5)$, $\psi(10)$, $s(5)$, $s(45)$. Then the τ for each field separately is conserved by all the interactions except the Yukawa one (10.23), and for the latter the total τ is seen to be conserved if the τ charges of the separate fields are chosen to be $\tau(\psi(10)) = 1$, $\tau(\psi(45^*)) = -3$ and $\tau(s(5)) = \tau(s(45)) = -2$, up to an overall normalization. Since the gauge fields A_μ and the scalars $s(24)$ are real, $\tau(A_\mu) = \tau(s(24)) = 0$. Now since the SSB $SU(5) \rightarrow S(U(3) \times U(2))$ is caused by $s(24)$ which is τ-neutral, it does not break the τ-symmetry, and thus at the $S(U(3) \times U(2))$ level there are *two* central $S(U(3) \times U(2))$-invariant) conserved charges, namely τ and the charge σ of (10.15). The second SSB, $S(U(3) \times U(2)) \rightarrow U(3)$, is caused by $s(5)$ and $s(45)$, which have (τ, σ) charges $(-2, 1)$. Hence in this breakdown τ and σ are not separately conserved, but the combination $\tau + 2\sigma$ *is* conserved. It is easy to verify that for the fermions $(\tau + 2\sigma) = 5(B - L)$. Thus $\frac{1}{5}(\tau + 2\sigma)$ is a natural extension of $B - L$ to the bosons, and is conserved. For the gauge fields and $s(24)$ scalars it is zero for all but the $SU(5)/S(U(3) \times U(2))$ (X-like) fields, for which it is $\pm\frac{2}{3}$, and for the $s(5)$ and $s(45)$ scalars it takes the values $(0, -\frac{2}{3})$ and $(0, 0, -\frac{2}{3}, -\frac{2}{3}, -\frac{2}{3}, \frac{2}{3}, -\frac{4}{3})$ for the multiplets in the $S(U(3) \times U(2))$ decomposition (10.19).

The fact that B and L are not separately conserved means that the proton can decay into leptons, and this result is by far the most spectacular prediction of $SU(5)$ (and of grand unification theories in general). To identify the bosonic fields through which it can decay one notes that the fundamental process must be $qqq \rightarrow l$, or, equivalently, $qq \rightarrow \bar{l}\bar{q}$. Hence the mediating field must have the quantum numbers of qq, and thus have $B - L \neq 0$. The bosons of this kind are evidently the X_μ gauge fields, for which the explicit couplings can be seen in (10.21), and the $s(5 + 45)$ scalars for which $(\tau + 2\sigma) \neq 0$ as listed above.

The decays via X_μ are evidently the analogues of the decays via the $W_\mu^\pm$ in the ordinary weak interactions, and so, to estimate the proton life-time one uses the standard Born-approximation formula $\tau_p \sim M^4 e^{-4} m^{-5}$ that is used for weak decays, but with $e = SU(5)$ gauge (or Yukawa) coupling constant, $m =$ proton mass, and $M =$ mass of mediating boson. If the present experimental bound $\tau_p > 10^{31}$ years is used, one finds that $M \gtrsim 10^{17}$ GeV (Langacker, 1981). Thus to keep the proton stable the masses of the mediating gauge fields X and mediating scalar fields must be very large. The closeness of the scale to the GUT scale found from the renormalization group considerations of the previous section is remarkable, and is one of the main reasons for the acceptance of the GUT idea.

However, a discrepancy of order 10^2 indicates that, while the general idea may be correct, the simple $SU(5)$ model as such, is probably not, and more detailed investigations (Langacker, 1981) confirm this view. Another piece of experimental evidence against the simple $SU(5)$ model is that (using the gauge sector) it predicts a cosmic baryon–photon asymmetry of order 10^{-16}, in disagreement with the experimental value of 10^{-10} (Ellis, 1981; Kolb and Turner, 1983). However, this evidence is less reliable because, unlike the proton-decay rate, the discrepancy could be removed by adjusting the (as yet undetermined) parameters in the scalar sector.

10.4   SO(10) GUT model

The simplest GUT group after $SU(5)$ is $SO(10)$, and because it explains some of the features of $SU(5)$ and $S(U(3) \times U(2))$ and illustrates some of the difficulties in symmetry breaking, it should perhaps be mentioned briefly. The group $SO(10)$ is a member of the class $SO(4n+2)$, which automatically satisfy the ABJ anomaly condition of section 5.4 because the groups are orthogonal, and which have strictly complex representations because their rank is odd (section 5.4). The group $SO(10)$ contains $SU(5)$ (as a subgroup of the group $U(5)$ which is the little group of the matrix $\mathrm{diag}(\mathbb{1}_5, -\mathbb{1}_5)$ with respect to conjugation) and it is the smallest member of the $SO(4n+2)$ class which contains $SU(5)$. The $SU(5)$ decomposition of the fundamental complex representation $\varDelta^+$, which is 16-dimensional, is $\varDelta^+ = 10 + 5^* + 1$, and thus, if the singlet is identified as the right-handed component of the neutrino, each $S(U(3) \times U(2))$ generation fits snugly into the $\varDelta^+$, which, unlike the $5^* + 10$ of $SU(5)$, is irreducible. The structure of the $SU(5)$ and $S(U(3) \times U(2))$ generations, and their freedom from anomalies, is then explained in terms of $SO(10)$.

The Yukawa couplings are, as usual, $(f \times f + f^* \times f^*)$ and since $(\varDelta^\pm \times \varDelta^\pm) = F_1 + F_3 + F_5^\pm = 10 + 120 + 126^\pm$ the scalars which cause the final breakdown $S(U(3) \times U(2)) \to U(3)$ should belong to the 10, 120 and $126^\pm$ and the scalars which cause the earlier breakdown, $SO(10) \to SU(5) \to S(U(3) \times U(2))$ say, should not. The assignment of the latter scalars is one of the problems of $SO(10)$. The natural assignments would be the adjoint and/or the symmetric tensor representations (the 45 and/or 54), but although these representations can easily generate little groups such as $U(5)$ and $U(3) \times U(2)$ (which have the same rank as $SO(10)$) from quartic potentials, they do not produce the little groups $SU(5)$ or $S(U(3) \times U(2))$. Hence one is forced to use more exotic combinations such as the $16 + \overline{16} + 45$ (Buccella, Ruegg and Savoy (1980)) the $10 + 54 + 126$ (Buccella, Cocco and Wetterich (1984) and the $126 + 210$ (Tuzzi (1985))).

10.5   Ground rules for GUT models

The $SU(5)$ and $SO(10)$ models by no means exhaust all the possibilities for grand unification and there have been many attempts to go beyond them, with both single-generation and multi-generation models (i.e. models which connect different $S(U(3) \times U(2))$ generations with so-called horizontal symmetry groups). In the case of the single-generation models the aim is to explain the $S(U(3) \times U(2))$ assignments and to derive the low-energy $S(U(3) \times U(2))$ phenomenology in a natural manner. In particular, the aim is to improve the prediction for the proton lifetime, to predict the existence of the top-quark (which rules out many previous models) and to obtain the experimental structure of the neutral weak $U(2)$ current J_μ^0, especially to obtain the experimental value $\sin^2 \theta_w \simeq \frac{2}{9}$ for the weak angle (or $\sin^2 \theta_w \approx \frac{3}{8}$ before renormalization). The aim of the multi-generation models is to extend these results to all the $S(U(3) \times U(2))$ generations simultaneously, and, if possible, to explain the existence of the three known generations, and obtain values for their mass spectra and for the components of the KM-matrix, which mixes the generations.

The number of proposed models, with and without horizontal symmetry, is so large that it would not be possible to describe them in detail here. So the plan will be to list first the general rules that are used for model building (these are actually constraints on the matter-field assignments) and then briefly describe the properties of the models, grouped according to the different type of grand unification group ($SU(n)$, exceptional, semi-simple etc.). In the present section the general ground rules are considered, and it will be convenient to consider the fermions and scalar fields separately.

Rules for fermions

The left-handed fermions and antifermions (i.e. ψ_L and $(\bar{\psi})_L = (\overline{\psi_R})$) are assigned to a representation f of the grand unification group G and their right-handed counterparts to the complex conjugate representation f^*. The main conditions on the representation f are:

(i)   The representation f should contain the $S(U(3) \times U(2))$ generation(s) and should agree with all the low-energy $S(U(3) \times U(2))$ phenomenology.

(ii) Since electroweak $U(2)$ violates parity and electrocolour $U(3)$ does not, the parts of f and f^* which survive at low energy should be inequivalent with respect to $U(2)$ and equivalent with respect to $U(3)$. This can be achieved either by letting the f and f^* be equivalent for the full GUT group and invoking a spontaneous breakdown of parity with respect to $U(2)$, or by letting f and f^* be strictly inequivalent (but equivalent with respect to the $U(3)$ subgroup) from the beginning. The second alternative is possible only for groups with strictly complex representations and it will be recalled that the only representations of this kind are the complex representations of $SU(n)$, the $\Delta^{\pm}$ representations of $SO(4n+2)$ and the fundamental 27-dimensional representation of $E(6)$.

(iii) The assignment f must satisfy the ABJ condition (7.26) for the absence of anomalies. It will be recalled that this problem arises only for the complex representations of $SU(n)$. For these representations, (which are commonly used) the condition comes into conflict with condition (ii), and the way out of this difficulty is to use *reducible* representations (such as the $5+10^*$ of $SU(5)$) chosen so that the anomalies from the different irreducible parts cancel.

(iv) The asymptotic freedom condition is not very stringent for fermions, so it begins to create problems only in the case of horizontal symmetry, where the inclusion of different generations means that the number of fermions becomes quite large.

(v) A useful first Ansatz that is often made is that the representation f of the original GUT group should contain only singlets and triplets with respect to the colour-$SU(3)$ subgroup (with equal number of 3s and 3*s on account of the parity condition (ii)). This Ansatz will be referred to as the GMRS condition, as it was first formulated by Gell-Mann, Ramond and Slansky (1978) who used it to classify the possible embeddings of $S(U(3) \times U(2))$. They found that there were essentially four classes of groups, namely two classes for which the leptons (l) and quarks (q) are transformed by different factors of the group i.e. groups of the form $G(l) \times G(q) \times U(1)$ or $G(l) \times G(q) \times G(\bar{q}) \times U(1)$ and two classes for which the leptons and quarks are transformed by the same factor, i.e. groups of the form $G_{l+q} \times U(1)$ and G_{l+q}. Here the $U(1)$ groups are used to distinguish quarks and leptons, and the electric charge is generally a combination of the generators from the different factors. A notable exception to GMRS Ansatz is the group $E(8)$, for which the Ansatz is not satisfied even in the fundamental representation.

Rules for scalars

(i) The *raison d'être* of the scalars is to produce a hierarchy of groups $G \to G' \to G'' \to S(U(3) \times U(2)) \to U(3)$ by spontaneous symmetry breakdown. Hence the first requirement is that they be assigned to representations that can produce the required groups G', G''... as little groups. A necessary condition for this is that the representation must contain singlets with respect to the various subgroups G', G''... and it often requires that the representation be reducible, a different irreducible component being responsible for each stage. The scale of each breakdown is determined by the value of $\overset{\circ}{\phi}$, the constant scalar singlet which determines the minimum of the potential at that stage.

(ii) In principle the successive spontaneous symmetry breakdowns in (i) should be implemented by a renormalizable (4th-degree polynomial) potential. In practice, the existence of such a potential is often assumed without explicit verification, partly because of the difficulty of constructing it, partly because so little is known experimentally about the scalar sector, and partly because the scalars are probably composite anyway.

(iii) The condition for asymptotic freedom is difficult to enforce for scalars, and is often ignored for the same reasons as the exact form of the potential.

(iv) Since the scalars generate fermion masses through the Yukawa couplings, some care must be taken in matching the scalar assignments to the fermion mass-spectrum. In particular, if f denotes the low-energy fermion representation and S_k the Yukawa representations, defined as the irreducible components in the expansion

$$f \times f + f^* \times f^* = \sum_k S_k,$$

then the representations of the scalars which cause the large-scale (say 10^{15} GeV) symmetry breakdown should *not* include the Yukawa (S_k) representations (otherwise these fermions acquire large masses). On the other hand the representations of the scalars which cause the low-energy (including electroweak) breakdowns *should* include at least some of the Yukawa representations (otherwise these fermions remain completely massless). Furthermore, the decomposition of these scalar representations with respect to electroweak $U(2)$ should contain the standard $U(2)$ scalar doublet, so at least some Yukawa representations with this property must be used.

Finally, for both the fermion and scalar assignments, the question as to whether the $B-L$ number should be conserved should be considered. A detailed discussion of this question has been given by Primakoff and Rosen (1981).

10.6 Models without horizontal symmetry

In this section a brief description will be given of the single-generation models that have been constructed following the rules of section 10.5. The models are collected according to their grand unification groups.

A. Models based on simple classical groups

These models have only one gauge coupling constant, have a high unification scale (say $M \gtrsim 10^{12}$ GeV) and are a natural choice, especially following on $SU(5)$ and $SO(10)$. One of their disadvantages is that the order of the group is arbitrary, or is limited only by asymptotic freedom. For the various Cartan classes one has:

$SU(n)$: These are the most natural groups following $S(U(3) \times U(2))$ and $SU(5)$ and have the advantage that most CUIRs are strictly complex. However the complex CUIRs have anomalies so one is forced to use reducible combinations of them. The most economical member $n = 5$ has already been discussed and its main drawback is that the proton decay rate is too large. For $n \geqslant 6$ the rate becomes more flexible but this has to be balanced against the fact that for $n \geqslant 6$ there are many branches of symmetry breaking to $S(U(3) \times U(2))$ (e.g. via $S(U(4) \times U(2))$ or $S(U(3) \times U(3))$) and these are not only arbitrary but for the most part lead to problems with the J_{μ}^0 data.

$SO(4n+2)$: These groups have the great advantage that they have no anomalies and at the same time they have strictly complex representations (generated by the primitive complex representations $\Delta^{\pm}$). The most economical case $SO(10)$ has already been discussed, and together with $E(6)$ (below) it is probably the most popular candidate for a (non-horizontal) GUT to date. When $n \geqslant 3$ there are roughly the same kind of advantages and disadvantages as for $SU(n)$ when $n \geqslant 6$.

$SO(4n)$, $SO(2n+1)$, $USp(2n)$: These groups have no anomalies, but they also have no strictly complex representations and thus a spontaneous breakdown of the real assignments must always be invoked. In practice $SO(2n+1)$ and $USp(2n)$ are rarely used and when $SO(4n)$ is used it is quickly broken down to $SO(4m+2) \times H$, where H is some other (say horizontal) group.

B. *Models based on exceptional groups*

These groups have no anomalies, a single coupling constant, and a high unification mass scale. They have been advocated particularly by Gürsey and his collaborators (see for example Gürsey and Sikivie (1977) and Koca (1981)) and their main attraction is that there is no arbitrariness in the order. The smallest exceptional group $G(2)$ is too small so that only the F and E groups need be considered.

$F(4)$, $E(7)$: These groups have no complex representations and for this reason and because of problems with the J_μ^0 data (e.g. $\sin^2 \theta \approx \frac{3}{4}$, no top quark, in many models) they have been largely abandoned.

$E(6)$: This is the only exceptional group with strictly complex representations (generated by the fundamental 27-dimensional complex representation) and with $SO(10)$ it is probably the most popular unification group. $E(6)$ has a number of branches to $S(U(3) \times U(2))$, for example via $SU(3)^3$ (which has problems with the J_μ^0 data) and via $S(O(10) \times O(2))$. The latter branch yields all the good $SO(10)$ and $SU(5)$ results and at the same time it has the flexibility to avoid the electron d-quark mass degeneracy $(m_e = m_d)$ which comes from using the 5 instead of the $5+45$ of $SU(5)$.

$E(8)$: This group has no complex representations and is one of the few groups which has no representations satisfying the GMRS Ansatz. Nevertheless $E(8)$ is popular because it is easily fitted to the J_μ^0 data and because it has good symmetry-breaking branches to $S(U(3) \times U(2))$, through $E(6)$ and $SO(16)$ for example. It has also been pointed out by Olive (1982) that it has the remarkable breakdown pattern

$$E(8) \to E(7) \to E(6) \to E(5) \equiv SO(10) \to E(4) \equiv SU(5)$$

$$\to E(3) \equiv S(U(3) \times U(2)),$$

modulo some central $U(1)$ groups, and that the discrete centres of these successive groups are just Z_n, $n = 0, 1, 2, ..., 6$. The group $E(8)$ has the added attraction that its fundamental and adjoint representations are the same, which makes it very suitable for supersymmetry, and it has no sixth-degree Casimir, which means that it can be used to construct an anomaly-free super-symmetric string theory (Green and Schwarz, 1984).

C.   Models based on semi-simple groups

At first sight these groups would appear to violate the first principle of grand unification since they allow more than one coupling constant. However, if the direct product consists of products of the same group with itself, e.g. $G = H^3 = H \times H \times H$, and there is a discrete symmetry permuting the factors the uniqueness of the coupling constant can be restored. The main advantage of these groups is that they allow the grand unification mass to be quite low, even as low as 1 TeV in some cases, without a corresponding reduction in the proton decay rate. In fact, the proton decay rate can be made arbitrarily small. The prototype of such models is the original grand-unification model based on $SU(4)^4$ (i.e. $SU(4)_l^w \times SU(4)_r^w \times SU(4)_l^s \times SU(4)_r^s$, where w, s, l, r denote weak, strong, left, right respectively) proposed by Pati and Salam (1973). The details of this model and modified versions of it have been considered by many authors (see for example Pati (1978) and Rajpoot (1979)).

10.7   Models with horizontal (and other) symmetries

The models considered in the previous section do not include horizontal symmetry groups, i.e. groups G_h which link together the different $S(U(3) \times (2))$ generations. To include horizontal groups one has to start from models based on much larger GUT groups. The same constraints as before apply to these larger models, and are handled in essentially the same way, but there are also some new constraints. First, the larger number of fermion fields which are involved in the different generations means that the conditions for asymptotic freedom due to the fermions become more stringent. Second, one must arrange the Yukawa couplings, so that the required generations, and only these, are left massless after the large mass-scale spontaneous breakdown(s). In other words one must choose the groups, representations and Yukawa couplings so that the observed low energy generations, and only these, 'survive' all but the final electroweak breakdown. From the phenomenological point of view there is the problem of obtaining the correct mass splitting between the generations and the correct generation-mixing KM-matrix, in particular the correct Cabibbo angle. There is also the problem of satisfying the cosmological constraints, notably the constraints on the number of different neutrinos, and therefore the number of generations. Finally there is the still-open question as to the mass scale at which the horizontal symmetry should break down. Should it precede the usual large-scale breakdown at 10^{15} GeV or follow it, taking place perhaps at a much lower scale such as 10^4 GeV?

Since the inclusion of horizontal symmetry represents some kind of final step in the unification process it is not surprising to find that simple groups are by far the most popular ones for grand unification groups which include horizontal symmetry. Among the simple groups the most popular are $SU(n)$, $SO(2n)$ and $E(8)$ (the other exceptional groups being too small) and among the $SU(n)$ groups in turn, the most popular is $SU(8)$ (see for example Fujimoto (1981), Kim and Roiesnel (1980), Baaklini (1980), Chaichan, Kolmakov and Nelipa (1982)). For all the $SU(n)$ groups the most popular representations for the fermion assignments are the l primitive ones (see for example Georgi (1979), Frampton (1979)). The $SO(2n)$ groups have the difficulty that the decomposition of the spinor representations Δ, $\Delta^{\pm}$ contain both the Δ^+ and Δ^- representations of any contained $SO(4n+2)$ and so some special mechanism has to be used to ensure that Δ^- generations do not survive at low energy. A recent revival of the $O(18)$ model due to Bagger and Dimopoulos (1984) illustrates this point and also the general situation for $SO(n)$ groups in a very readable manner.

As mentioned before, the exceptional group $E(8)$ has good branches leading to $S(U(3) \times (2))$, and among those useful for horizontal symmetry are $SU(3)_h \times E(6)$ and $SU(3)_h \times SU(5)$. For this reason, and because of its supersymmetric properties, $E(8)$ remains a popular candidate. For a review see Stech (1980).

A general problem with horizontal symmetry is that if the horizontal symmetry group is continuous then its spontaneous breakdown will entail the existence of either Goldstone scalars or the gauge fields which absorb them according to the Higgs mechanism, and neither set of fields has actually been observed. To escape this problem it has been proposed that the horizontal symmetry group should be finite and a number of models with such finite horizontal symmetry groups have been constructed (see for example Ecker (1984), Mohapatra (1985)). However, the models are rather ad hoc, and in general it is not easy to obtain little groups with non-trivial discrete components from the spontaneous breakdown of continuous groups (at least using conventional representations).

Finally it might be mentioned that there have been attempts to incorporate other groups such as the Peccei–Quinn axial $U(1)$ group, and the so-called 'technicolour' groups into the grand unified scheme. Here by the PQ axial $U(1)$ groups are meant groups that have been introduced in order to rotate away the CP-violating effects that come from the presence of instantons in non-abelian gauge theories. (For a recent discussion and a proposal to incorporate an axial $U(1)$ in the horizontal symmetry group see Davidson, Nair and Wali, 1984). By 'technicolour' is meant a very high energy group whose purpose is to generate a 'dynamical' symmetry breakdown i.e. a group whose fermions and gauge fields are supposed to form scalar condensates which take the place of the usual scalar fields, thus eliminating the need for fundamental scalar fields altogether. For a review see Kaul (1983).

10.8   Pros and cons of grand unification

Although grand unification is a natural continuation of $S(U(3) \times U(2))$ theory, it is by no means inevitable, and in this section the arguments for and against it are assembled.

The arguments for it may be divided into the general arguments for gauge theory and arguments based on the details of $S(U(3) \times U(2))$. The general arguments are

(1)   The success of electroweak theory.
(2)   The universality of the gauge principle, as evidenced by $S(U(3) \times U(2))$ and gravitation.
(3)   The probable triviality of pure scalar theory (Fröhlich, 1982).
(4)   The good intrinsic properties of gauge theories, notably their geometrical origin, their asymptotic freedom, and their presumed connection with confinement.
(5)   The agreement between grand unification and cosmology.

The arguments from the details of $S(U(3) \times U(2))$ theory are

(a)   The detailed success (to date) of $S(U(3) \times U(2))$ itself.
(b)   The possibility GUT offers of overcoming the main weakness of $S(U(3) \times U(2))$, namely the existence of three independent coupling constants.
(c)   The quantization of the electric charge for simple G.
(d)   The existence of a scale ($\sim 10^{15}$ GeV) at which the three $S(U(3) \times U(2))$ constants merge into one (section 10.2).
(e)   The fact that this scale falls roughly between the bound set by proton decay and the Planck mass.
(f)   The existence of a previously tested mechanism (SSB for $U(2)$) to explain the breakdown of G to $S(U(3) \times U(2))$.

(g) The fact that all known physical states belong to $S(U(3) \times U(2))$ rather than $SU(3) \times SU(2) \times U(1)$. This means that the $SU(3)$, $SU(2)$ and $U(1)$ subgroups are already correlated globally, and is the situation that arises naturally in SSB.

(h) The neat way in which the known fermion generations fit into the fundamental representations of $SU(5)$ and $SO(10)$ suggests unification at least to the level of these groups.

Some arguments against grand unification are

(A) No fully satisfactory GUT group G has yet been found.

(B) The energy range $10^2 - 10^{15}$ GeV is so enormous that it is difficult to believe that present ideas could be extrapolated over it without any major change.

(C) The enormity of this range also leads to the so-called gauge-hierarchy problem. This problem will be discussed in section 12.4 and is perhaps the most serious problem confronting grand unification.

(D) It is difficult to explain the existence of $n \geqslant 3$ identical generations using only Lie groups.

(E) The large number of 'elementary' fermions required (at least six quarks and two leptons in each generation) suggests that, even if GUT is correct, it is not fundamental.

(F) Even for $U(2)$ the scalar sector has not been seen directly. Hence the GUT scalar sector is an extrapolation of a scheme which is itself only tentative.

(G) GUTs always have instanton solutions, and hence have the CP-problems due to the degenerate instanton-connected θ-vacua (Callan, Dashen and Gross, 1976; Quinn and Peccei, 1979; Dine, Fischler and Srednicki, 1981; Davidson, Nair and Wali, 1984).

(H) For most GUT models the scalar fields belong to reducible and/or large group representations, and thus the unification in the gauge-fermion sector is achieved only at the expense of introducing a large number of free parameters in the scalar sector (as $S(U(3) \times U(2)) \to SU(5)$ for example, $g_1, g_2, g_3 \to e$, but the number of parameters in the scalar potential increases from 2 to 5 (chapter 9)). Thus, unless there is an independent principle (e.g. supersymmetry) to determine it, the scalar sector is not really unified.

The last four objections would be met if GUTs were not regarded as a fundamental description but merely as a useful semi-phenomenological one (like the Landau theory of superconductivity) and this may very well be their role.

Finally it should be pointed out that there is great difficulty in testing GUTs experimentally. Proton decay is the only direct prediction that is measurable, but while its observation would be spectacular evidence in favour of GUTs, failure to observe it is not strong evidence against them. Apart from proton decay there is only the indirect evidence from cosmology and future very high energy (TeV) experiments.

Exercises

10.1. Give an argument to show that in the approximation that the scalar and fermion loops are neglected the value of the common gauge coupling $g^2(M) = F^{-1}(M)$ is just the $U(1)$ coupling g_1^2 ($\approx 2e_0^2$ for $c^2 \approx \frac{3}{8}$) and deduce that this is a lower bound for the true value.

10.2. Check that the full form for the Yukawa interaction (10.23) in tensor notation is

$$g\epsilon_{abcde}\bar{\psi}_{ab}\psi_{cd}S_e + h\bar{\psi}^{ab}\psi_b S_a + k\bar{\psi}^{ab}\psi_c S_{ab}^c + \text{herm. conj.}$$

where S_a and S_{bc}^a are the 5- and 45-dimensional representations $S(5)$ and $S(45)$, with

$$S_{ab}^a = 0, \quad S_{bc}^a + S_{cb}^a = 0.$$

Show that if 1, 2, 3 are the colour indices and 5 the neutral electroweak index then the colour and electromagnetic invariant vacuum values of $S(5)$ and $S(45)$ must be $\overset{\circ}{S}_5 (= m$ say) and $\overset{\circ}{S}_{45}^4 = -3\overset{\circ}{S}_{\alpha5}^\alpha (= -3\mu$ say) $\alpha = 1, 2, 3$, and hence that the mass spectrum for the first generation of fermions is

$$m_e = mh - 3\mu k, \quad m_d = (mh + \mu k), \quad m_u = mg \quad (m_\nu = 0).$$

10.3. Show that the first three primitive representations of $SU(8)$ have the $SU(5) \times SU(3)$ decompositions

$$8 = (1, 3) + (5, 1), \quad 28 = (10, 1) + (5, 3) + (1, \bar{3})$$
$$56 = (10, 3) + (5, \bar{3}) + (10, 1) + (1, 1)$$

and hence that $SU(8)$ has the *a priori* possibility of being used for horizontal symmetry containing three generations. From the formula for anomalies in the primitive representations of $SU(n)$, show that the anomalies are (1, 2, 3) respectively, and hence that $8_R + 28_R + 56_L$ will satisfy the anomaly condition and reduce to $5_R + 10_L$ for each generation. Using the formula for asymptotic freedom (with $s = 0$) show that $SU(8)$ and the subgroups $SU(5)$ and $SU(3)$ are asymptotically free but $SU(2)$ is asymptotically neutral (Fujimoto, 1981).

11
Orbit structure

11.1 Introduction

In chapter 4 it was pointed out that the intrinsic properties of spontaneous symmetry breaking are contained in the orbital structure, where the orbits are defined as the sets of vectors $U(g)\overset{\circ}{\phi}$, where $\overset{\circ}{\phi}$ is a minimal point of the potential. The purpose of this chapter is to consider this orbital structure in a more detailed and systematic manner. The results obtained may not be sufficient, or even necessary, to determine the breakdown pattern in specific cases, but they give an overall view and often help to shorten the labour. Unfortunately, even for irreducible representations the number of cases for which complete results are available is rather limited, indeed consists mainly of fundamental and second-rank tensor representations. However, from the known results some general patterns, and also some interesting conjectures have emerged.

As discussed in chapter 8, each orbit has an intrinsic little group H. However, since the number of orbits in a representation is infinite (inequivalent norms, for example, imply inequivalent orbits) and the number of little groups is finite, it is clear that the correspondence is not one–one. On the other hand, orbits with the same little group have similar physical properties (the same number of massive gauge fields, for example) so it is convenient to collect them into classes. The word used to describe such a class is a *stratum*. The strata are in one–one correspondence with the little groups and thus each stratum contains *all* the orbits in a representation which have the same little group. For example, in the defining, n-dimensional representation of $SO(n)$, there are two strata, the trivial $SO(n)$ stratum $\phi = 0$, and the continuous $SO(n-1)$-stratum with orbits $(\phi, \phi) = s$, $0 < s < \infty$.

Because orbits are transitive ($x \in 0(y)$ if, and only if, there exists a $g \in G$ such that $x = U(g)y$) and $U(h)y = y$ where $h \in H$ is any element of the little group H, it is clear that the points in the orbits are in one–one correspondence with the points in the cosets G/H, and thus the topology of one provides a natural topology for the other. For example, orbits for

128

which $H = 1$ have the topology of the group itself, and for $SU(2)$ the orbits $SU(2)/U(1)$ have the natural topology of the 2-sphere S_2. The coset topology is evidently the same for all orbits in the same stratum.

In principle the study of little groups requires a knowledge of all the (closed) subgroups K of a given group, and the representations for which they are little groups (for which there are K-singlets). This is obviously a very extensive problem and is by no means completely solved at the present time. Much work has been done on it, especially by Dynkin, but to present even the known results would be beyond the scope of this monograph. Accordingly, the reader is referred to the recent book by Cahn (1984) and to the review articles by Slansky (1981) and by MacKay and Patera (1981) in which excellent surveys, directed at physicists and based on Dynkin's work, are given. In these reviews particular attention is paid to the following groups:

maximal subgroups; i.e. those not contained in larger subgroups
maximal little groups; i.e. those not contained in larger *little* groups
minimal little groups; i.e. those not containing any smaller *little* groups

as these are of special importance for spontaneous symmetry breaking. Note that the minimal little group need not be $H = 1$ (it often is not) and that the little groups are representation dependent by definition.

An awkward problem in practice is that the representations are not always irreducible. Whether this represents an incompleteness in the unification scheme, or is due to the composite nature of the scalar fields is not known, but at any rate it complicates the algebra, and most of the known results are limited to the irreducible case.

The plan of the next sections will be to first build up some intuition about orbits and little groups by means of examples, and then present some general results. (The trivial orbits $\overset{\circ}{\phi} = 0$ will be neglected.)

11.2 $SU(2)$ orbits

Even for $SU(2)$ the orbital structure is non-trivial and of physical importance. For the fundamental representation $j = \frac{1}{2}$ every 2-vector ψ can be brought by an $SU(2)$ transformation to the form $(0, |\psi|)$ and no $SU(2)$ transformation leaves this form invariant. Hence there is only one stratum, its little group is $H = 1$, and the orbits are parametrized by the norm $|\psi|$. For the adjoint representation $j = 1$, the base vectors of the representation may be represented by Pauli matrices σ_i, $i = 1, 2, 3$, on which the group acts by conjugation. Since every vector $\mathbf{a} \cdot \boldsymbol{\sigma}$ may be conjugated into $|a|\sigma_3$

by an $SU(2)$ transformation and this vector is left invariant by the $U(1)$ group $\exp i\alpha\sigma_3$, there is again only one stratum, and the orbits are parametrized by the norm. But the little group is $U(1)$.

For $j \geqslant \frac{3}{2}$ the situation is more complicated. $U(1)$ is the only proper continuous subgroup of $SU(2)$ and from the result for $j = 1$ it can always be conjugated into $\exp i\alpha\sigma_3$. Hence if it is a little group there exists a vector ψ in the representation for which $\sigma_3\psi = 0$. It is well known from angular momentum theory that this is never the case when $2j$ is odd and occurs for exactly one vector (modulo the norm) when $2j$ is even. Thus for $2j$ odd (spinor representations of $SO(3)$) the strata have little groups which are at most discrete, and since the dimension of the representation is $2(2j+1)$ and that of $SU(2)$ is 3, the orbits are parametrized by $4j-1$ parameters. (Thus only for $j = \frac{1}{2}$ does the norm give a complete parametrization.) For $2j$ even (tensor representations of $SO(3)$) on the other hand, there is one stratum with $U(1)$ as little group, and its orbits are parametrized by the norm. The remaining strata have at most finite little groups, and their orbits are parametrized by the remaining $(2j+1)-3 = 2(j-1)$ parameters.

The simplest example of the latter case is $j = 2$, where the vectors in the representation may be thought of as real symmetric traceless 3×3 matrices on which the group acts by conjugation with $SO(3)$. Since such matrices can be diagonalized (and the eigenvalues ordered) by conjugation with $SO(3)$, and no distinct set of eigenvalues $\lambda_1 \geqslant \lambda_2 \geqslant \lambda_3$, where $\lambda_1 + \lambda_2 + \lambda_3 = 0$ can be conjugated into another, the orbits may be parametrized by these sets of eigenvalues. The $U(1)$ stratum corresponds to the degenerate case when two eigenvalues are equal, and the other stratum to the non-degenerate case. Note that the little group for the non-degenerate case is the finite abelian group of order 4, $\mathrm{diag}\,(\epsilon_1, \epsilon_2, \epsilon_3)$, $\epsilon_i^2 = 1, \epsilon_1\epsilon_2\epsilon_3 = 1$ in $SO(3)$, and that this group is covered by the non-abelian quaternionic group of order 8, $(\pm 1, \pm i\sigma)$ in $SU(2)$.

11.3 Orbits for first- and second-rank tensors of $SO(n)$

The fundamental (n-dimensional) and second-rank tensor representations of $SO(n)$ can be handled by the methods used for $SU(2)$. First, any vector ϕ in the fundamental representation can be brought to the form $(0, 0, \ldots, 0, |\phi|)$ by an $SO(n)$ transformation, and since the little group of this vector is $SO(n-1)$, there is just one stratum, with little group $SO(n-1)$ and the orbits are parametrized by the norm.

For the symmetric traceless ($\frac{1}{2}n(n+1)$ dimensional) second-rank tensor representation the vectors may be represented by symmetric traceless matrices T and the group action by conjugation. Thus it is a generalization of the $j = 2$ representation of $SU(2)$ discussed above. The matrices can be diagonalized (and the eigenvalues ordered) by $SO(n)$ conjugation, and no distinct set of eigenvalues $\lambda_1 \geqslant \lambda_2 \geqslant \ldots \geqslant \lambda_{n-1} \geqslant \lambda_n$ where $\lambda_1 + \lambda_2 + \ldots + \lambda_n = 0$, can be conjugated into another. Thus, once more, these sets of eigenvalues exactly parametrize the orbits. The little groups H are determined by the degeneracies, thus $H =$ finite, finite $\wedge$ $SO(2)$, finite $\wedge$ $SO(3)$, finite $\wedge$ $S(O(2) \times O(2))$ etc. for all eigenvalues different, two eigenvalues equal, three eigenvalues equal, two pairs of eigenvalues equal, etc. Since the degeneracies may be characterized by the partitions of n, the little groups and hence the strata, may be characterized by the partitions of n (omitting the trivial partition because of the trace condition).

For the antisymmetric ($\frac{1}{2}n(n-1)$-dimensional) second-rank tensors, the procedure is analogous except that since T is antisymmetric it can be conjugated only to the 2×2 block-diagonal form $T = \sigma \times \operatorname{diag}(\lambda_1 \ldots \lambda_l)$ where $\sigma = \begin{pmatrix} 0 & 1 \\ -1 & 0 \end{pmatrix}$, l is the rank ($2l = n$ or $n-1$) and an extra diagonal zero must be added to T in the case of odd n ($2l = n-1$). Then the orbits are again characterized by the distinct ordered sets of eigenvalues $\lambda_1 \geqslant \lambda_2 \geqslant \ldots \geqslant \lambda_l$, and the strata by the partitions of l. Note, however, that there are only l eigenvalues and there is no trace condition. The minimal and maximal little groups are $SO(2)^l$ and $U(l)$ respectively.

11.4   Orbits for first- and second-rank tensors of $SU(n)$

For the defining and adjoint representations of $SU(n)$ the situation is analogous to the defining and symmetric tensor representations of $SO(n)$. Any vector ψ in the defining representation can be brought to the form $(0, 0 \ldots 0, |\psi|)$ by an $SU(n)$ transformation. Hence there is one stratum, and the orbits are parametrized by the norm. The only new feature is that the little group is $SU(n-1)$. Similarly, any traceless hermitian matrix in the adjoint representation can be $SU(n)$ conjugated to diagonal form with the (real) eigenvalues ordered, and distinct sets of ordered eigenvalues are not conjugate. Thus the orbits may be parametrized by the sets of ordered eigenvalues $\lambda_1 \geqslant \lambda_2 \geqslant \ldots \geqslant \lambda_{n-1} \geqslant \lambda_n$, where $\lambda_1 + \lambda_2 + \ldots + \lambda_n = 0$, and the strata by the degeneracies of these eigenvalues, i.e. the non-trivial partitions of n. The corresponding little groups are evidently $S(U(n_1) \times U(n_2) \times \ldots \times U(n_p))$. Note that the minimal little group, corresponding to the maximal partition (into n units), is $U(1)^{n-1}$ and the maximal little groups, corresponding to the minimal partitions $n = p+q$, are $S(U(p) \times U(q))$.

A new situation occurs for the symmetric (and antisymmetric) second-rank tensor representations of $SU(n)$ because the tensors ψ are not hermitian (or anti-hermitian) and the group action is $\psi \to g^t \psi g$, which is not a conjugation, (except for the $SO(n)$ subgroups). The strategy then is to write the tensor as $\psi = u+iv$ where u, v are real and have the same symmetry properties as ψ, and to diagonalize in the following three steps: First, by using the fact that the group action on $\psi^\dagger \psi$ *is* by conjugation, $\psi^\dagger \psi \to g^\dagger (\psi^\dagger \psi) g$, the quantity $\psi^\dagger \psi$ is made diagonal and the (real non-negative) eigenvalues ordered. In this basis u and v commute since $[u,v] = \text{Im } \psi^\dagger \psi = 0$. Second, an $SO(n)$ transformation is used to make the commuting real-symmetric (or real-antisymmetric) matrices u and v diagonal (or 2×2 block diagonal). Finally, a diagnoal $U(1)^{n-1}$ transformation is used to make the diagonal (or 2×2 block diagonal) elements of $u+iv$ real and non-negative (modulo an overall phase factor if there are no zero eigenvalues, i.e. if ψ is non-singular). The end-result is that ψ is of the form $\psi = \text{diag}(\lambda_1, \lambda_2, ..., \lambda_q, 0...0)$, $q \leqslant n$ for ψ symmetric, $\psi = \text{diag}(\lambda_1 \sigma, \lambda_2 \sigma, ..., \lambda_q \sigma, 0, ..., 0)$, $q \leqslant \frac{1}{2}n$, $\sigma = \left(\begin{smallmatrix} 0 & 1 \\ -1 & 0 \end{smallmatrix} \right)$ for ψ antisymmetric, and $\lambda_1 \geqslant \lambda_2 \geqslant ... \geqslant \lambda_q > 0$ in both cases. Since ψ cannot be further reduced by $SU(n)$ transformations, this result shows that the orbits may be parametrized by the positive spectrum $\lambda_1 \geqslant \lambda_2 ... \geqslant \lambda_q$, together with the matrix rank of ψ (number of zero eigenvalues) and one overall phase (if there are no zero eigenvalues). The little groups are therefore of the form $H = S(H(q) \times U(n-q))$, where $H(q)$ is the little group of $O(q)$ determined by the multiplicity structure of the positive spectrum (just as in the $SO(n)$ cases of the previous section).

11.5 Symmetric algebras; adjoint of $SU(n)$

The variety of strata found for the second-rank tensor representations of $SO(n)$ and $SU(n)$ show that, unlike the fundamental representation, they are by no means isotropic with respect to the group. A convenient basis-independent manner to describe the anisotropy, or what might be called the geometry, of these representations has been introduced by Biedenharn (1963) and by Michel and Radicati (1971, 1973) for the adjoint representation of $SU(n)$, and is easily extended to the combined (symmetric + antisymmetric) tensor representations of $SO(n)$, and to other cases.

The idea will be sketched for the adjoint representation of $SU(n)$. If A and B denote matrices in the fundamental $(n \times n)$ representation of the $SU(n)$ Lie algebra, then the anticommutator $\{A, B\}$ is not, in general, an element of the Lie algebra because of the trace condition, but if one defines the symmetric product

$$A \vee B = \tfrac{1}{2}\{A, B\} - \frac{1}{2n} \operatorname{tr}\{A, B\}, \tag{11.1}$$

then this product is an element. Thus the product $A \vee B$ defines a *symmetric* algebra on the space of the Lie algebra, i.e. the adjoint representation space. It is interesting to compare it with the antisymmetric algebra $A \wedge B \equiv [A, B]$ defined by the Lie algebra itself. Both algebras are $SU(n)$-invariant because

$$U(g) A \circ B U^{-1}(g) = U(g) A U^{-1}(g) \circ U(g) B U^{-1}(g), \tag{11.2}$$

where $\circ = {}_\wedge$ or ${}_\vee$, and, in analogy to (3.7)

$$(A \vee B, C) = (C \vee A, B) = (B \vee C, A), \tag{11.3}$$

where brackets denote the Cartan inner product. However, the analogue of (11.3) for four elements A, B, C, D does not hold, and the operation ${}_\vee$, in contrast to ${}_\wedge$, does not obey the Leibnitz rule for derivation (see exercise 11.2).

The symmetric algebra can be used to characterize the degeneracies of the eigenvalues in an invariant way. For example a matrix with $p+1$ distinct eigenvalues satisfies an equation of $(p+1)$th degree in the symmetric algebra, and is called *p*-potent. In particular, matrices with only two distinct eigenvalues satisfy the equation

$$I \vee I = kI, \tag{11.4}$$

where

$$k = \frac{p-q}{n} \quad \text{for } I = I(p, q) = \frac{1}{n}\begin{pmatrix} p\mathbb{1}_q & \\ & -q\mathbb{1}_p \end{pmatrix}, \quad \left(\operatorname{tr} I^2 = \frac{pq}{n}\right),$$

and are called *idempotents*. Idempotents are important for SSB because they are the matrices which minimize the scalar potential for the adjoint representation (section 12.2).

Another important set of elements are the *charges Q_i*, $i = 1, \ldots, l$ defined (for any Lie algebra) as the elements of the Cartan which combine with the elements $E_{\pm i}$ associated with primitive roots α_i to form l $SU(2)$ subalgebras $\{E_i, E_{-i}, Q_i\}$, and which for $SU(n)$ are just the elements $\operatorname{diag}(1, -1, 0, 0, 0, \ldots)$, $\operatorname{diag}(0, 1, -1, 0, 0, \ldots)$ etc. They satisfy the cubic equation $Q_i^3 = Q_i$ and have inner products $(Q_i, Q_j) = C_{ij}$ where C_{ij} is the

Cartan matrix (2.39). They play an important role in the theory of magnetic monopoles (Goddard and Olive, 1978; Weinberg, 1980) where the quantization condition $\exp(4\pi_i Q) = 1$ implies that (up to conjugation) $Q = \Sigma n_i Q_i$, where the n_i are integers.

The charges are not idempotents but are related to idempotents in two important ways. First their squares in the symmetric algebra are idempotents

$$Q_i \vee Q_i = I(n-2, 2) \tag{11.5}$$

with the 2×2 unit matrix opposite the non-zero part of Q_i. Second their duals P_i, defined as

$$(P_i, Q_j) = \delta_{ij}, \tag{11.6}$$

are just the idempotents $I(p, q)$ for $p = 1, 2, ..., n-1$ (with inner products $(P_i, P_j) = (C^{-1})_{ij}$). The duality of charges and idempotents mirrors the duality of primitive roots and primitive weights and is of importance for extending the conventional electric–magnetic duality to non-abelian gauge theories (Goddard and Olive, 1978), for monopole-induced proton decay (Callan, 1982; Rubakov, 1982) and for colour breaking by monopoles (Balachandran *et al.*, 1984).

The extension of the symmetric algebra to the combined symmetric + antisymmetric representations of $SO(n)$ is obvious, but it can also be extended to other representations. For example, for the $(0100...010)$ representation of $SU(n)$ carried by tensors of the form S_B^A where $S_A^A = d_B^A(a) S_A^B = 0$, $A, B = 1...\frac{1}{2}n(n-1)$ the symmetric algebra can be written as

$$(S \vee T)_B^A = U_B^A - \sum_{a=1}^{n^2-1} d_B^A(a)\, d_D^C(a)\, U_C^D - \frac{2}{n(n-1)} \delta_B^A U_C^C, \tag{11.7}$$

where

$$U_B^A = \tfrac{1}{2}(S_C^A T_B^C + T_C^A S_B^C), \quad d_B^A(a)\, d_A^B(b) = \delta_{ab}$$

and the $d_B^A(a)$ are Clebsch–Gordon coefficients for the coupling of the indices A, B to the index a of the adjoint representation.

11.6 General orbital structure

The results on the orbit structure of representations found in the previous examples are special cases of more general results which will now be summarized. The proofs may be found in the papers by Michel (1971, 1972, 1980) and references quoted therein.

The first general result is that in any continuous unitary irreducible representation (CUIR) of a compact Lie group there can only be a *finite* number of strata (little groups). Furthermore the little groups can be partially ordered (where ordering is defined as $H_i > H_j$ if H_i contains a G-conjugate of H_j). With this ordering the minimal little group, H_0 say, is unique. Thus the general little group structure has an ordering something like

where arrow denotes inclusion, and the structure for the adjoint representation of $SU(5)$ is given on the right as an example. Note that the maximal little groups (those towards which no arrows point) are not necessarily unique.

In general the smaller the little group the larger the number of orbits in its stratum. This can be seen by noting that if $H(v_0 + \epsilon v)$ is the little group of the vectors $v_0 + \epsilon v$ in the ϵ-neighbourhood of the vector v_0 then, by continuity, $H(v_0 + \epsilon v) \leqslant H(v_0)$. In particular it can be shown that the stratum corresponding to the minimal little group H_0 is always so large that it is open dense in the space of orbits, and for this reason it is often called the *generic* stratum. A good example of this result is furnished by the adjoint representation of $SU(n)$ for which it reduces to the statement that any real traceless diagonal $n \times n$ matrix can be written as the limit of a sequence of real traceless diagonal matrices with n distinct eigenvalues. At the other extreme from the generic strata are strata in which there is only one orbit, or only a discrete number of orbits, modulo the norm. The orbits in such strata, which are obviously isolated within the strata, are called *critical* orbits and their little groups are very often maximal. In particular, if there is only one orbit in a stratum modulo the norm, i.e. if the little group has only one singlet in the representation, then it is easy to see that the little group must be maximal by definition. All the maximal little groups of the first- and second-rank tensor representations discussed in sections 11.3 and 11.4 are of this kind, i.e. have unique singlets.

For a given CUIR, a subgroup H of a group G will be a little group if, and only if, the decomposition of the CUIR with respect to H contains at least one H-invariant vector (an H-singlet). Since this is not always the case (for example the adjoint of $SU(3)$ splits into the $5 + 3$, but no 1, with respect to the real $SO(3)$ subgroup) it is clear that a given subgroup is not necessarily a little group. On the other hand there is a theorem due to Mostow (1958) which shows that for any given (closed) subgroup H of a compact connected Lie group G, there exists at least one representation (not necessarily irreducible) for which H is a little group. One way in which

Mostow's result often comes into play is when a given subgroup H becomes a little group because it is the intersection of the little groups for two (or more) irreducible components of a reducible representation. For example an $SU(3)$ little group of $SU(5)$ could occur as the intersection of the $SU(4)$ little groups for the 4 and 4* components of the reducible 4+4* representation. Indeed if a given subgroup occurs as a little group for some irreducible representation, but the irreducible representation is too large to handle with comfort, then it is often more convenient to obtain it as the little group for a smaller, reducible, representation. For example it may be more convenient to obtain the trivial little group $H = 1$ of $SO(3)$ from the 3+3 rather than from the irreducible 7-dimensional representation.

An interesting question is whether there is a loss of rank in a breakdown $G \rightarrow H$, and if so, how much. An upper bound is given by $\Delta l = l - l_0$ where l is the rank of G and l_0 the rank of the minimal little group H_0, since, for all other little groups H, $l_0 < l_H < l$. For the first- and second-rank tensor representations of $SO(n)$ and $SU(n)$ considered in the previous sections one sees that:

$\Delta l = 1$ for the first-rank tensors (defining representations)
$\Delta l = 0$ for the adjoints of $SU(n)$ and $SO(n)$ and the anti-symmetric
   tensor of $SU(n)$
$\Delta l = l$ for the symmetric tensors of $SU(n)$ and $SO(n)$.

Thus there may be no loss of rank or a complete loss of rank in going from G to H_0. In general, if the vector $\overset{\circ}{\phi}$ for which H_0 is the stability group is an eigenvector of the Cartan algebra

$$U(H_i)\overset{\circ}{\phi} = h_i \overset{\circ}{\phi}, \quad i = 1, ..., l \tag{11.8}$$

then, if $h_i = 0$ the whole Cartan algebra is in the little group and $\Delta l = 0$, and if any h_i, h_1 say, is not zero, then the $(l-1)$-dimensional subalgebra $h_1 H_i - h_i H_1$ is in the little group, and $\Delta l = 1$. Physically important examples of $\Delta l = 1$ occur in the spontaneous breakdown of electroweak $U(2)$ into electric $U(1)$, and of electroweak–strong $S(U(3) \times U(2))$ into electro–strong $U(3)$, (chapter 10), and a problem with grand unified $SO(10)$ is to produce $\Delta l = 1$ so that it breaks to $SU(5)$ (section 10.4). So far as the author is aware there is no systematic study of the Δl problem.

Finally it is often useful to use the concept of the *little space R* of a little group H, defined as the subspace of all vectors left invariant by H ($H\phi = \phi \Leftrightarrow \phi \in R$). In particular, if one considers a hierarchy, or completely ordered chain, $H_0 < H_1 < H_2 ... < H_m < G$, in the partially ordered set of little groups, then it is easy to see that one has the ordering $R_0 > R_1 > R_2 ... > R_m > 0$. Furthermore the inclusions are strict i.e.

$\dim R_k - \dim R_{k+1} \geqslant 1$, because $R_k = R_{k+1}$ would imply $H_{k+1}\phi = \phi$ for all $H_k\phi = \phi$, in which case H_k would not be a little group by definition. This generalizes the earlier statement that $\dim R = 1$ implies that H is maximal.

11.7   Orbits and invariants

In this section the group invariants $I(\phi)$ of representations are considered, as the invariants are related to the orbital structure and are important for potential theory (next chapter). By definition the invariants satisfy the condition

$$I(U(g)\phi) = I(\phi), \qquad (11.9)$$

where ϕ are the vectors in the representation space, and for the CUIRs of compact Lie groups they are monomials, whose degree will be denoted by a subindex k. Thus $I = I_k$. For example, for the symmetric tensor representation of $SO(n)$, the invariants are $I_k(\phi) = \operatorname{tr}\phi^k$, $k = 2, ..., n$. Of course, there may be many invariants of a given degree. For example, for the (2, 2) or 27-dimensional representation of $SU(3)$ carried by the 8×8 traceless real symmetric tensor T_{ab}, where $d_{ab}^c T_{ab} = 0$, there are two independent third-degree invariants, namely $\operatorname{tr} T^3$ and $f_{bc}^a f_{jk}^i T_{ai} T_{bj} T_{ck}$ where the fs are the structure constants.

For the simple compact Lie groups there are no first-degree invariants ($I_1 = 0$), but there is always a second-degree invariant and (by Schurs lemma) it is unique up to a constant in the irreducible case. The second-order invariant $I_2(\phi)$ can be expressed as an inner product (ϕ, ϕ) and used as a norm to make the other invariants dimensionless, $I_k \to J_k = I_k/(I_2)^{\frac{1}{2}k}$. It is isotropic in the group (does not distinguish any particular vectors) and is actually invariant with respect to the larger group $U(d)$ (or $SO(d)$ if it is real) where d is the dimension of the CUIR. The higher-degree invariants do not, in general, have these properties.

The importance of the invariants for the orbital structure is that they are constant along the orbits (by definition) but vary in the orbit space. In fact they vary to such an extent that they 'separate' the orbits (Schwarz, 1975) and thus can be used to parametrize the orbits in an invariant manner. It will be recalled, for example, that the orbits for the adjoint representation of $SU(n)$ are parametrized by the ordered sets of eigenvalues $\lambda_1 \geqslant \lambda_2 \geqslant ... \geqslant \lambda_n$, $\lambda_1 + \lambda_2 + ... + \lambda_n = 0$, and since

$$I_k = \operatorname{tr}\phi^k = \sum_a \lambda_a^k, \quad k = 2, ..., n, \qquad (11.10)$$

the I_k give an equivalent, but manifestly invariant, parametrization. Since the strata are characterized by degeneracies in the eigenvalues, they are characterized by algebraic relations among the I_k, and this is a general feature of the invariants for any group or representation.

A more useful description of the relationship between orbits and invariants may be obtained by considering the extrema of the invariants

$$\frac{\partial I_k(\phi)}{\partial \phi} = 0, \qquad (11.11)$$

(modulo $I_1 = 0$ and possibly $I_2 =$ constant) or, more generally, their Morse theory (Michel, 1980; Houston and Sen, 1984; Witten, 1982) because the extrema define little groups and thus help to characterize the strata. For example, for the adjoint representation of $SU(n)$, where the $I_k(\phi)$ are as shown in (11.10) equation (11.11) reduces to

$$\phi^{k-1} = \text{constant}, \quad (k \geqslant 3), \qquad (11.12)$$

which shows that, for $3 \leqslant k < n$, ϕ has at most $k-1$ distinct eigenvalues and so the little group is at least as large as $S(U(m_1) \times U(m_2) \times \ldots \times U(m_{k-1}))$ where $n = m_1 + \ldots + m_{k-1}$ is a $(k-1)$ partition of n. In particular, for $I_3(\phi)$ the little group must be a maximal little group $S(U(p) \times U(q))$, $p+q = n$. Note that the minimal little group $H_0 = U(1)^l$ never occurs as a little group of (11.12) because, for $3 \leqslant k \leqslant n$, at least two eigenvalues must be equal. The special invariant $I_2(\phi)$ is excluded because its only extremum is $\phi = 0$ (whose little group is $SU(n)$ itself).

For general groups and CUIRs the situation is rather similar to that just described for the adjoint representation of $SU(n)$. The only extremum of I_2 is the G-invariant orbit $\phi = 0$, and the extrema of the other invariants lie on orbits with little groups between G and H_0, but are neither G nor H_0. Thus the extrema never lie in the dense generic stratum. At least one extremum of each I_k lies on a maximal orbit, and for the lowest order invariants all the extrema lie on maximal orbits. Conversely, in each CUIR there is at least one orbit on which *all* the I_k (except I_2) have extrema. These are just the critical orbits discussed earlier.

So far only the extrema of the invariants have been considered, but, of course, the maxima and minima are particularly important, especially for potential theory. The local minima (and maxima) are characterized by the fact that the *Hessian* $\partial^2 V / \partial \phi_\alpha \partial \phi_\beta$ is a positive (negative) matrix, (though not positive (negative) definite on account of the Goldstone theorem) and in many cases the absolute minima are easily found once the local ones are known. However, the Hessian condition is more difficult to analyse than (11.11) since it is only an inequality, and in many cases it is easier to proceed directly from (11.11). For example, for the adjoint representation of $SO(n)$, the extremal equation for $J_4(\phi) = (\operatorname{tr} \phi^4)/(\operatorname{tr} \phi^2)^2$ is evidently

$$\phi^3 \operatorname{tr} \phi^2 = \phi \operatorname{tr} \phi^4 \quad \text{or} \quad \phi^4 = s\phi^2, \tag{11.13}$$

where $s = J_4(\operatorname{tr} \phi^2)$, and by substituting this result back into J_4 one finds that at any extremum $J_4 = (2m)^{-1}$ where $2m$ is the dimension of the space on which ϕ^2 is not zero. Thus (for $\operatorname{tr} \phi^2 \neq 0$) the minima and maxima occur for $2m = n, n-1$ for n even and odd, and for $2m = 2$ respectively. The little groups are easily seen to be $Sp(m)$ and $S(O(2) \times O(n-2))$, respectively.

The little groups of the maxima and minima in the last example are maximal little groups, and this has been found to be a general result – although the intermediate extrema (saddle-points) of invariants tend to have non-maximal little groups, the little groups for the maxima and the minima of the invariants tend to be maximal.

Exercises

11.1. By considering the weight diagram of the $(2, 2)$ or 27-dimensional representation of $SU(3)$, show that this representation contains a (unique) singlet with respect to the subgroups $U(2)$, $SO(3)$ and $W \wedge (U(1) \times U(1))$ where W is the Weyl group, so that these are maximal little groups.

11.2. Show that in the symmetric algebra of the adjoint representation of $SU(n)$

$$A \vee (B \vee X) - B \vee (A \vee X) = n[[B, A], X] + 2B(A, X) - 2A(B, X)$$

(Michel and Radicati, 1973).

11.3. Show that for any CUIR Σ of a compact simple Lie group the number of independent invariants is

$$\dim \Sigma - \dim G + \dim H_0,$$

where H_0 is the minimal little group (and $\dim \Sigma$ is the number of *real* dimensions). Verify this for the fundamental and second-rank tensor representations of $SO(n)$ and $SU(n)$.

11.4. Show that the maxima of $\operatorname{tr} \phi^4/(\operatorname{tr} \phi^2)^2$ and $\operatorname{tr} \phi^3/(\operatorname{tr} \phi^2)^{\frac{3}{2}}$ in the adjoint representation of $SU(n)$ have little groups $U(n-1)$ ($\operatorname{tr} \phi^2 \neq 0$).

11.5. Show that, apart from the fundamental CUIR of each of the simple compact groups (except $E(8)$), and its conjugate for $SU(n)$ and $E(6)$, the only CUIRs with $\dim R < \dim A$ are the symmetric 2-tensors of $SU(n)$, $n \geq 3$, the antisymmetric 2-tensors of $SU(n)$, $n \geq 4$, and of $Sp(2l)$, $l \geq 2$, the totally antisymmetric 3-tensors of $SU(6)$, $SU(7)$, $SU(8)$ and $Sp(6)$ and the fundamental spinorial representations Δ, $\Delta^\pm$ of $SO(7)$, ..., $SO(14)$.

12
The scalar potential

12.1 General considerations

The scalar potential $V(\phi)$ plays a crucial role in unified gauge theory because it governs the spontaneous symmetry breakdown (SSB). In the ideal case where $V(\phi)$ is given, the problem is to determine the SSB patterns, i.e. the possible little groups, but in practice the problem is often the reverse: given a pattern, to construct a renormalizable potential that will produce it. In either case, what is required is the relationship between the potential and the patterns, and it is the general form of this relationship that will be considered in this chapter. The potential V is, of course, a group invariant, $V(\phi) = V(U(g)\,\phi)$, and in the renormalizable (fourth-degree polynomial) case it is therefore a polynomial in the invariants $I_k(\phi)$, $k = 2, 3, 4$ of chapter 11. Thus it is of the very restricted form,

$$V(\phi) = V_4(\phi) + V_3(\phi) + V_2(\phi), \tag{12.1}$$

where $V_k(\phi)$ denotes a linear combination of the independent invariants of degree k. The local minima are given by

$$\frac{\partial V(\phi)}{\partial \phi_\alpha} = 0, \qquad\qquad \frac{\partial^2 V}{\partial \phi_\alpha \partial \phi_\beta} \geq 0. \tag{12.2}$$

As was seen in chapter 11 the $V_3(\phi)$ and $V_4(\phi)$ are not necessarily isotropic in the representation. This means that in general there will be many directions of breaking, i.e. many patterns, and that the case $\lambda(I_2(\phi)^2 - c^2)^2$ of (8.3), where the SSB is caused only by the functional dependence of V on I_2, and there is only one pattern, is a very special (and rather trivial) case. Most of the recent literature on symmetry breaking patterns (see further reading section) has been devoted to investigating the patterns permitted by V_3 and V_4. As there are no systematic methods available, the methods which are used in those investigations are, perhaps, best illustrated by specific examples.

Before going on to the examples there are some general points that might be mentioned. First, the general tendency of the invariants I_k to have maxima and minima on maximal orbits carries over to the potentials. This empirical tendency is so strong that it even led to a conjecture (Michel, 1980) that if the potential is renormalizable (fourth-degree) and the representation is irreducible then the absolute minimum of the potential always lies on a maximal orbit. It is now known that the conjecture is not universally true (Abud *et al.* 1984, Burzlaff, Murphy and O'Raifeartaigh 1985), but it is true in many relevant cases, e.g. for all the fundamental and second-rank tensor representations of $SO(n)$ and $SU(n)$ (see section 12.2) and, in general, it expresses the maximizing tendency very well. Even the counter-examples are only slightly less than maximal. As mentioned in section 11.6 the maximality of the little group is closely related to the question as to whether the minimum vector $\overset{\circ}{\phi}$ is a unique singlet for the little group.

The second point is that the simple dependence of the potential on the invariants $I_k(\phi)$ suggests that these invariants could be used as the variables in place of ϕ, and this approach, which has been systematically exploited by Kim (1982), has the advantage that it is group invariant and gives some geometrical insight. But it requires a knowledge of the domains of the $I_k(\phi)$ and that can be quite complicated. In general the generic orbits correspond to the interior of the domains (which is open dense) the other little groups to the boundaries, and the maximal little groups to singular boundary points such as cusps.

Thirdly the results of Mostow and Schwarz mentioned in sections 11.6 and 11.7 respectively may be used to show that if there are no restrictions on the reducibility of the representation or the functional form of the potential then any subgroup K of a given group may be the stability group for a potential minimum. To see this let R be a Mostow representation, i.e. a representation containing a K-singlet, $\overset{\circ}{\phi}$ say. Since R contains only a finite number of irreducible components, and for each component there is only a finite number of invariants (finite number of strata) it follows from Schwarz's result that it is possible to find a function $f(R, I(\phi))$ of the invariants that separates the orbits in R. Let c be the value of $f(R, I(\phi))$ on the orbit of $\overset{\circ}{\phi}$. Then the potential $V(\phi) = (f(R, I(\phi)) - c)^2$ has the subgroup K, and only this subgroup, as stability group. In many grand unification models a stronger result is assumed, namely that even for fourth-degree polynomials an arbitrary subgroup K may be a stability group provided the representation of the scalar fields is sufficiently reducible. This seems plausible, but the author is not aware of any formal proof. In any case, what the Mostow–Schwarz result shows is that the usual problems encountered in determining symmetry-breaking patterns come mainly from the extraneous conditions that are imposed, namely a certain degree of irreducibility (to reduce the number of parameters) and the restriction to fourth-degree polynomials (for renormalizability).

Finally it should be mentioned that since for each irreducible component ϕ of the scalar field there is only one second-order invariant $I_2(\phi)$, the extremal equation (12.2) takes the form

$$\frac{\partial I_4(\phi)}{\partial \phi} + \frac{\partial I_3(\phi)}{\partial \phi} + m\phi = 0, \qquad (12.3)$$

where $\tfrac{1}{2}m$ is the coefficient of ϕ^2 in the potential, and while, in principle this is an equation to be solved for $\overset{\circ}{\phi}$ in terms of m and the other parameters, in practice (when $\overset{\circ}{\phi} \neq 0$) it is often convenient to regard it as an equation for m in terms of $\overset{\circ}{\phi}$. This is reasonable from the physical point of view because $\overset{\circ}{\phi}$ has the more immediate physical significance (recall that $e^2(\sigma_a\overset{\circ}{\phi}, \sigma_b\overset{\circ}{\phi})$ is the mass matrix for the gauge fields) and it means that (12.3) can be solved at once to give

$$m = -(4I_4(\overset{\circ}{\phi}) + 3I_3(\overset{\circ}{\phi}))/(\overset{\circ}{\phi}, \overset{\circ}{\phi}). \qquad (12.4)$$

This result can be quite useful, especially as $I_3(\overset{\circ}{\phi})$ or $I_4(\overset{\circ}{\phi})$ can be eliminated from it by using the extremal equation (8.28).

12.2 Examples for irreducible representations

It was seen in chapter 11 that the defining representation of $SO(n)$ or $SU(n)$ contains only one non-trivial stratum of orbits. Consequently the only algebraically independent invariants for these representations are $I_2(\phi) = (\phi, \phi)$ and the only renormalizable potentials are of the rather trivial kind

$$V(\phi) = \lambda((\phi, \phi) - c^2)^2, \qquad (12.5)$$

which have already been considered in section 8.1.

For the second-rank tensors of these groups there is more variety. For the symmetric and antisymmetric tensor representations of $SU(n)$, and the adjoint representation of $SO(n)$, the potential is of the form

$$V(\phi) = \frac{h}{4!}(\operatorname{tr}\phi^\dagger\phi)^2 + \frac{g}{4!}\operatorname{tr}(\phi^\dagger\phi)^2 + \frac{m}{2}\operatorname{tr}(\phi^\dagger\phi), \qquad (12.6)$$

where ϕ is symmetric, antisymmetric, and real-antisymmetric respectively. These potentials have been treated in detail by Li (1974). For the adjoint representation of $SU(n)$ and the symmetric tensor representation of $SO(n)$ the potential may also have a cubic term, i.e.

$$V(\phi) = \frac{h}{4!}(\operatorname{tr}\phi^2)^2 + \frac{g}{4!}\operatorname{tr}(\phi^4) + \frac{f}{3!}\operatorname{tr}\phi^3 + \frac{m}{2}\operatorname{tr}\phi^2, \quad (\operatorname{tr}\phi = 0) \quad (12.7)$$

where ϕ is hermitian for $SU(n)$ and real-symmetric for $SO(n)$. Of course, if required, (12.7) could be reduced to (12.6) by invoking reflexion invariance in ϕ (and because of the reflexion symmetry the reduction would be stable with respect to radiative corrections). The case (12.7) with $f \neq 0$, $g \neq 0$ will be considered as an illustration of a non-trivial symmetry breakdown. The parameters in (12.7) are free except for the fact that $V(\phi)$ must be bounded below, a condition on h and g that emerges from the subsequent analysis.

The first result for (12.7) is obtained from the extremal equation

$$\frac{g}{3!}\phi^3 + \frac{f}{2}\phi^2 + \left(\frac{h}{3!}\operatorname{tr}\phi^2 + m\right)\phi = \text{constant}. \tag{12.8}$$

This shows that at any extremum ϕ has at most three distinct eigenvalues, i.e.

$$\phi = \operatorname{diag}(x\mathbb{1}_p, y\mathbb{1}_q, z\mathbb{1}_r), \tag{12.9}$$

where $px + qy + rz = 0$. The little group H is clearly $S(U(p) \times U(q) \times U(r))$ for $SU(n)$ (or $S(O(p) \times O(q) \times O(r))$ for $SO(n)$), unless two of the eigenvalues are equal, in which case it reduces to $S(U(p) \times U(q))$ (or $S(O(p) \times O(q))$). Unless two eigenvalues are equal, H is not maximal, and the representation contains two H-singlets, namely, ϕ itself, and a multiple of $\operatorname{diag}((y-z)/p, (z-x)/q, (x-y)/r)$.

The second important result for (12.7) is that if the extremum is a minimum (even a local minimum), then two of the eigenvalues x, y, z are indeed equal. Then ϕ can be written as

$$\phi = x\operatorname{diag}(p\mathbb{1}_q, -q\mathbb{1}_p), \tag{12.10}$$

the little group H is $S(U(p) \times U(q))$ or $S(O(p) \times O(q))$ and ϕ is the only H-singlet. Thus both the maximality conjecture of section 12.1 and the unique-singlet result quoted in 11.6 hold in this case. There are a number of proofs of this result available (Ruegg, 1980; Murphy, O'Raifeartaigh, 1983) but the following one, the idea for which was communicated to me by P. Mithra (1984) is perhaps the simplest and the most susceptible to generalization. Consider any $K = S(U(3) \times U(n-3))$ subgroup of $SU(n)$ (or $K = S(O(3) \times O(n-3))$ subgroup of $SO(n)$) which is simultaneously block-diagonal with ϕ. Then the matrix ϕ has a unique K-decomposition $\phi = \lambda + \mu + \omega$ where λ is $SU(n-3)$ $(SO(n-3))$ invariant μ is $SU(3)$ $(SO(3))$ invariant, and ω is K-invariant. Since λ is 3×3 and traceless, one has

$$\operatorname{tr}\lambda^4 = \tfrac{1}{2}(\operatorname{tr}\lambda^2)^2, \quad -6^{-\frac{1}{2}}(\operatorname{tr}\lambda^2)^{\frac{3}{2}} \leqslant \operatorname{tr}\lambda^3 \leqslant 6^{-\frac{1}{2}}(\operatorname{tr}\lambda^2)^{\frac{3}{2}}. \tag{12.11}$$

Hence if $V(\phi)$ is expanded in terms of λ it becomes

$$V(\lambda, \omega, \mu) = \frac{(f + g\omega)}{3!} \operatorname{tr} \lambda^3 + U(\operatorname{tr} \lambda^2, \mu, \omega). \qquad (12.12)$$

The Hessian (second derivative) of V with respect to the variables ω and $t = \operatorname{tr} \lambda^3$ is

$$H(t, \omega) = \begin{bmatrix} U_{\omega\omega} & \frac{1}{6}g \\ \frac{1}{6}g & 0 \end{bmatrix}, \qquad (12.13)$$

and since $\det H$ is then $-(\frac{1}{6}g)^2$ one sees that V can have no minimum in the range where t is a variable. Hence the minima must occur on the boundary where, from (12.9), $6(\operatorname{tr} \lambda^3)^2 = (\operatorname{tr} \lambda^2)^3$. But this is just the condition that λ have only two distinct eigenvalues. On the other hand, since λ can be *any* 3×3 submatrix of the diagonal matrix ϕ, i.e. corresponds to *any* three eigenvalues of ϕ, it follows that ϕ can have only two distinct eigenvalues, as required. Note that the proof essentially reduces the problem of showing that $SU(n)$ $(SO(n))$ is maximally broken to showing that a small subgroup, namely $SU(3)$ $(SO(3))$, is maximally broken.

The result that the matrix which minimizes the potential can have only two distinct eigenvalues means that it satisfies the idempotent condition of section 11.5 and hence to consider the symmetry-breaking patterns in more detail it is convenient to rewrite the potential (12.7) as

$$V(\phi) = \frac{1}{4}g \operatorname{tr} [(\phi \vee \phi) + 2(f/g) \phi]^2 + \frac{1}{4}k[(\operatorname{tr} \phi^2) + c]^2, \qquad (12.14)$$

where the new constants k and c are given by

$$k = h + g/n, \quad \text{and} \quad kc = (2/g)(3mg - f^2). \qquad (12.15)$$

Here $\phi \vee \phi$ denotes the symmetric product of section 11.5, a constant $\frac{1}{4}kc^2$ has been dropped and it is assumed that $g \neq 0$. (If $g = 0$ the cubic invariant $\operatorname{tr} \phi^3$ is the only non-isotropic one and it is easily seen from the result of section 11.7 that if there is a spontaneous breakdown then $U(n-1)$ (or $O(n-1)$) is the only possible little group.) The advantage of the form (12.14) of the potential is that it makes the results which have been found for the patterns (Ruegg, 1980; Murphy, O'Raifeartaigh and Yamada, 1983) intuitively evident. The main results are:

(i)   The condition that V be bounded below is $(npq) k + (p-q)^2 g > 0$ for all $p + q = n$.

(ii) There is a spontaneous breakdown if, and only if, $kc > 0$ (not $m < 0$).

(iii) If $g < 0$ the first term in (12.14) is maximal at the potential minimum and since $\phi \vee \phi$ is proportional to $(p-q)(pq)^{-\frac{1}{2}}$ this will happen when $q-p$ is maximal. Then the little group is $U(n-1)$ (or $O(n-1)$). So for $g \leqslant 0$ there is only one pattern.

(iv) If $g > 0$ the first term in (12.14) is minimal at the potential minimum and one sees at once that in this case all maximal little groups are possible. In fact, if $c > 0$, $k > 0$ and $2f/g$ is chosen to be $(p-q)(npq)^{-\frac{1}{2}}$ then V reaches its absolute minimum (zero) if, and only if, ϕ is conjugate to $c(pqn)^{-\frac{1}{2}} \operatorname{diag}(pI_q, -qI_p)$ which has little group $S(U(p) \times U(q))$.

(v) The parameter that determines the pattern for $g > 0$ is cg^2/f^2.

(vi) As might be expected from (iv) the coefficients f and g (> 0) of the non-isotropic invariants pull in opposite directions, preferring $|p-q|$ to be as large and as small as possible, respectively. In particular if $g = 0$ the pattern is $(p,q) = (n-1, 1)$, and if $f = 0$ (the potential is reflexion-invariant) then the pattern is $(\frac{1}{2}n, \frac{1}{2}n)$ for even n and $(\frac{1}{2}(n+1), \frac{1}{2}(n-1))$ for odd n.

After the spontaneous breakdown there are four mass multiplets, namely the Goldstone fields, which are off-diagonal with respect to the $p \times p$ and $q \times q$ blocks, the polar field π, which is a multiple of $\operatorname{diag}(p\mathbb{1}_q, -q\mathbb{1}_p)$, and the fields which lie within the $p \times p$ and $q \times q$ blocks and are traceless (unless p or $q = 1$, when they vanish). The latter fields are evidently the adjoint (or symmetric tensor) representations of the subgroups $SU(p)$, $SU(q)$ (or $SO(p)$, $SO(q)$).

The mass of the polar field may be obtained from the first and second variations of $V(\phi)$ with respect to the scale of ϕ, as discussed in section 8.4 and, choosing to eliminate the second-order invariant I_2, i.e. the parameter m, in accordance with (12.4), one obtains

$$m_\pi^2(\operatorname{tr} \mathring{\phi}^2) = \frac{1}{6}[h(\operatorname{tr} \mathring{\phi}^2)^2 + g \operatorname{tr}(\mathring{\phi})^4] + \frac{1}{4}f \operatorname{tr} \mathring{\phi}^3. \tag{12.16}$$

The other two massive fields, ϵ say, commute with $\mathring{\phi}$ and are trace-orthogonal to it. Hence the second variation of $V(\phi)$ at $\mathring{\phi}$ with respect to them can be read off from (12.7) and is

$$\Delta V = \frac{1}{12}h(\operatorname{tr} \mathring{\phi}^2) \operatorname{tr} \epsilon^2 + \frac{1}{4}g \operatorname{tr}(\mathring{\phi}^2 \epsilon^2) + \frac{1}{2}f \operatorname{tr}(\mathring{\phi} \epsilon^2) + \frac{1}{2}m \operatorname{tr} \epsilon^2. \tag{12.17}$$

Hence, using (12.8) to eliminate m, one finds that the masses $M(\rho)$ are given by

$$M(\rho)^2(\operatorname{tr} \mathring{\phi}^2) = \frac{1}{12}g[2 \operatorname{tr}(\rho \mathring{\phi}^2) \operatorname{tr} \mathring{\phi}^2 - \operatorname{tr} \mathring{\phi}^4] + \frac{1}{4}f[2 \operatorname{tr}(\rho \mathring{\phi}) \operatorname{tr} \mathring{\phi}^2 - \operatorname{tr} \mathring{\phi}^3], \tag{12.18}$$

where ρ is the projection on the $(p \times p)$ or $(q \times q)$ sub-blocks.

12.3   Example for a reducible representation

As a final example let us consider the case of a reducible representation, namely the sum of the defining and adjoint representation of $SU(n)$. This particular reducible representation is chosen because it is used for the grand unified $SU(5)$ model ($n = 5$), and because it is useful to illustrate the gauge-hierarchy problem (next section).

Let the fields in the defining (n-dimensional) and adjoint ((n^2-1)-dimensional) representations of $SU(n)$ be denoted by η and ϕ respectively. If, for simplicity, one requires also the reflexion invariance $\phi \to -\phi$, then the most general $SU(n)$-invariant fourth-degree potential for η and ϕ is

$$V(\phi, \eta) = \left\{ \frac{\lambda}{4!}(\eta, \eta)^2 + \frac{\mu}{2!}(\eta, \eta) \right\} + \left\{ \frac{h}{4!}(\operatorname{tr}\phi^2)^2 + \frac{g}{4!}(\operatorname{tr}\phi^4) + \frac{m}{2}\operatorname{tr}\phi^2 \right\}$$
$$+ \{\epsilon(\eta, \eta)\operatorname{tr}\phi^2 + \omega(\eta, \phi^2\eta)\}. \quad (12.19)$$

(Without reflexion invariance there would be two cubic terms, one in the ϕ-sector and one in the interaction sector.) It will be assumed that $g \neq 0$, since otherwise the symmetry group of the ϕ-sector increases to $SO(n^2-1)$, and that $\omega \neq 0$ since otherwise there is a separate $SU(n)$-symmetry group for η and ϕ. In any case, the interaction with gauge (or fermion) fields, which are only $SU(n)$-invariant, would produce g and ω terms in the radiative corrections and thus it is natural to assume that g and ω are at least of order e^2 where e is the gauge (or Yukawa) coupling constant.

The first step in minimizing (12.19) is to write down the extremal equation for $\eta^\dagger$, namely

$$\phi^2\eta = c^2\eta, \quad (12.20)$$

where

$$c^2 = (-\omega)^{-1}\left[\frac{\mu}{2} + \frac{2\lambda}{4!}(\eta, \eta) + \epsilon(\operatorname{tr}\phi^2)\right].$$

Equation (12.20) shows that (at an extremum) η is an eigenvector of ϕ^2 and hence that ϕ^2 and η can be brought simultaneously to the block-form

$$\phi^2 = \operatorname{diag}(\phi_1^2, \phi_2^2, ..., \phi_q^2, c^2), \quad \eta = (0, 0, ..., 0, \eta_c), \quad (12.21)$$

where the blocks are not necessarily one dimensional. Then except in the special case where c^2 has multiplicity greater than one, and both $\pm c$ occur, (12.21) can be further reduced to

$$\phi = \operatorname{diag}(\phi_1, \phi_2, ..., \phi_{n-1}, c), \quad \eta = (0, 0, ..., 0, |\eta|), \quad (12.22)$$

where now each block, except possibly the c-block, has multiplicity one, but, of course, ϕ may have degeneracies $\phi_i = \phi_j$. [Even in the special case, it is easy to see that (12.21) may be reduced to

$$\phi = \mathrm{diag}\,(\phi_1, \phi_2, ..., \phi_{n-2}, -c, c), \quad \eta = (0, 0, ..., \sin\delta, \cos\delta)\,|\,\eta\,|,$$
$$(12.23)$$

for some angle δ, where each block now has multiplicity one and some of ϕ_i may be equal to $\pm c$. Then, since the potential does not depend on δ, and $\delta = 0$ is a special case of (12.22), it is sufficient for minimization purposes to consider (12.22) (keeping in mind, of course, that if the eigenvalue c^2 is degenerate (12.23) is another possibility for the fields).]

Once η and ϕ are of the form (12.22) it is natural to separate ϕ into a part $x\,\mathrm{diag}\,(-1, -1, ..., n-1)/(n(n-1))^{\frac{1}{2}}$ representing c, and a trace-orthogonal part $\phi_\perp$, which belongs to the adjoint representation of the little group $SU(n-1)$ of η. Then the potential becomes

$$V(\phi, \eta, x) = \left\{ \frac{\lambda}{4!}(\eta, \eta)^2 + \frac{\mu}{2}(\eta, \eta) \right\} + \left\{ \frac{\bar h}{4!}x^4 + \frac{m}{2}x^2 \right\} + \bar\varepsilon(\eta, \eta)x^2$$
$$+ \left\{ \frac{h}{4!}(\mathrm{tr}\,\phi_\perp)^2 + \frac{g}{4!}\,\mathrm{tr}\,\phi_\perp^4 + \frac{f}{3!}\,\mathrm{tr}\,\phi_\perp^3 + \frac{\tilde m}{2}\,\mathrm{tr}\,\phi_\perp^2 \right\}, \quad (12.24)$$

where the new parameters $\bar h$, $\bar\varepsilon$, f and $\tilde m$ are

$$\bar h = h + g\left(\frac{n^2 - 3n + 3}{n(n-1)}\right), \quad \bar\varepsilon = \varepsilon + \omega\frac{n-1}{n}, \quad f = \frac{-gx}{\sqrt{n(n-1)}},$$
$$\tilde m = m + 2\varepsilon\,|\,\eta\,|^2 + \left(\frac{h}{3!} + \frac{g}{2n(n-1)}\right)x^2. \quad (12.25)$$

For the field $\phi_\perp$ the problem is then reduced to that of the previous section, and thus at a minimum $\phi_\perp$ must be of the form $\phi_\perp = y\,\mathrm{diag}\,(p\mathbb{1}_q, -q\mathbb{1}_p)$ where $p + q = n - 1$. Furthermore, the little group $H = S(U(p) \times U(q))$ for $\phi_\perp$ is the little group for the whole system. By inserting the $\phi_\perp$ of this form into (12.24) it is reduced to a potential $V(x, \eta, y)$ for the three H-singlets x, y, η (and the pattern parameter $k = p - q$). The minima of this potential are easily found (Buccella, Ruegg and Savoy, 1980; Murphy, O'Raifeartaigh and Yamada, 1984) but as the details are somewhat tedious only the results for the patterns will be presented here. They are

$$g < 0 \begin{cases} \omega > 0 : & H = U(n-2) \\ \omega < 0 : & H = SU(n-1), \quad (\phi_\perp = 0) \end{cases}$$

$$g > 0 \begin{cases} \omega > 0 \begin{cases} : & H = S(U(\tfrac{1}{2}(n-1)) \times U(\tfrac{1}{2}(n-1))), \ n \ \text{odd}, \\ : & H = S(U(\tfrac{1}{2}(n-2)) \times U(\tfrac{1}{2}n)), \ n \ \text{even}. \end{cases} \\ \omega > 0 : & \text{All patterns } S(U(p) \times U(q)), \quad p = 0, 1, ..., n-1. \end{cases}$$

These results are understandable from the analyses of the previous section, where it was seen that for $g < 0$, $k \to 0$, n, and for $g > 0$, $k \to \frac{1}{2}n$, $\frac{1}{2}(n-1)$ unless f intervenes. One finds that actually the domain of the parameters for which all patterns other than the first three can be obtained is quite restricted. In particular it vanishes in the large n limit, which can be understood by noting that in this limit the adjoint representation dominates and $f \to 0$ in (12.25).

Let us now consider the masses which are generated by the spontaneous breakdown. For this purpose it is convenient to write $\mathring{\phi}$ as diag $(a\mathbb{1}_p, b\mathbb{1}_q, c)$, where $pa+qb+c$ is zero, to write $|\mathring{\eta}|$ as t, and to let σ and τ be the generators of the central $(S(U(p) \times U(q))$-invariant) $U(1)$ groups parallel to the direction $\mathring{\phi}$ and orthogonal to it.

For the gauge fields the mass formula, from (8.8), is

$$M = e^2[2 \operatorname{tr} A_\mu^2 \mathring{\phi}^2 - 2 \operatorname{tr}(A_\mu \mathring{\phi})^2 + \mathring{\eta}^\dagger A^2 \mathring{\eta}], \qquad (12.26)$$

and from this it is easy to see that the squares of the masses of the gauge fields are described by the $n \times n$ block-matrix

$$M^2(A_\mu) = \begin{bmatrix} 0 & 2e^2(a-b)^2 & 2e^2(a-c)^2+e^2t^2 \\ 2e^2(a-b)^2 & 0 & 2e^2(b-c)^2+e^2t^2 \\ 2e^2(a-c)^2+e^2t^2 & 2e^2(b-c)^2+e^2t^2 & 0 \end{bmatrix},$$

$$(12.27)$$

together with

$$M^2(A_\tau) = 0, \quad M^2(A_\sigma) = e^2t^2,$$

where the masses are entered in the positions of their fields. The scalar fields may be written as the corresponding tensors

$$\phi - \mathring{\phi} = \begin{bmatrix} \phi(p) & \theta & \rho_p \\ \theta & \phi(q) & \rho_q \\ \rho_p & \rho_q & 0 \end{bmatrix} \text{ plus } \phi(\sigma), \phi(\tau), \quad \eta - \mathring{\eta} = \begin{bmatrix} \eta_p \\ \eta_q \\ u+iv \end{bmatrix}, \quad (12.28)$$

where $\phi(p), \phi(q)$ are traceless, and u, v, $\phi(\sigma)$, $\phi(\tau)$ are real singlets. The computation of their masses from the potential is straightforward but tedious so only the main results will be given. First the Goldstone fields are the θ-multiplet, the singlet v, and the rotated vectors

$$\rho(\alpha) = \rho_p \cos\alpha + \eta_p \sin\alpha, \quad \rho(\beta) = \rho_q \cos\beta + \eta_q \sin\beta, \quad (12.29)$$

where $\tan\alpha = t/(c-a)$ and $\tan\beta = t/(c-b)$. The magnitude of the angles α and β can be understood from the fact that the Goldstone fields (12.29) are the ones that combine with the gauge fields of mass squared $e^2((c-a)^2+t^2)$ and $e^2((c-b)^2+t^2)$ in the Higgs mechanism, and the general structure of the Goldstone fields may be understood from its discussion in the next section.

The massive scalar fields are then the $\phi(p)$ and $\phi(q)$ multiplets, for which

$$M^2(\phi(p)) = 2g(a-b)(2a+b), \quad M^2(\phi(q)) = 2g(b-a)(2b+a),$$

$$(12.30)$$

the vector combinations $\rho_\perp(\alpha)$ and $\rho_\perp(\beta)$ orthogonal to the Goldstone fields (12.29), for which

$$M^2(\alpha) = 2\omega\left(\frac{a+c}{a-c}\right)[(a-c)^2+t^2], \quad M^2(\beta) = 2\omega\left(\frac{b+c}{b-c}\right)[(b-c)^2+t^2],$$

$$(12.31)$$

and the real singlets $\phi(\sigma)$, $\phi(\tau)$ and u. The latter three fields mix (the mass matrix is completely off-diagonal) and the simplest way to deal with them is to return to the 3-field potential $V(x, y, \eta)$ above, express it in terms of $\phi(\sigma)$, $\phi(\tau)$, u and proceed from there. The only simple result is that the expectation value of the mass squared of the polar field $\phi(\sigma)$ is $\frac{1}{2}(m+\mu)$ (see section 8.4). Note that the factors g and ω in (12.30) and (12.31) may be understood from the fact that the $\phi(p)$, $\phi(q)$ fields become Goldstone fields for $SO(n^2-1)$ when $g=0$, and the $\rho_\perp(\alpha)$, $\rho_\perp(\beta)$ fields become Goldstone fields for $SU(n) \times SU(n)$ when $\omega = 0$.

The special case $n = 5$ ($p = 3$, $q = 1$) of the above model is the scalar sector used for $SU(5)$ grand unification, and in that case the parameters $(a-b, a-c)$ and $(b-c, t)$ are of very different orders of magnitude, namely, 10^{15} GeV and 10^2 GeV respectively. The (B, L)-violating fields which cause proton decay as discussed in section 10.3, are then the gauge fields with mass squares

$$2e^2(a-b)^2 \quad \text{and} \quad e^2[2(a-c)^2+t^2] \approx 2e^2(a-c)^2, \qquad (12.32)$$

in (12.27), and the scalar fields $\rho_\perp(\alpha)$, with mass squares

$$2\omega\left(\frac{a+c}{a-c}\right)[(a-c)^2+t^2] \approx 2\omega(a^2-c^2), \qquad (12.33)$$

in (12.31). In order to agree with the observed bounds on the proton lifetime, the mass squares (12.32) and (12.33) must be very large ($\approx 10^{30}$ GeV^2) and in order that the scalar-mediated decay rate should not exceed the gauge-field-mediated rate, one must evidently have $\omega \gtrsim e^2$. It is interesting that this is just the lower bound on ω that is set by the radiative corrections as discussed at the beginning of this section.

12.4 Goldstone structure for a two-stage symmetry breakdown

In this section the Goldstone structure for a general two-stage symmetry breakdown will be considered. This may help to clarify the results found for the model of the previous section (and for other grand-unification models) and will also be useful in the discussion of the gauge-hierarchy problem in the next section. As in the model of the previous section ϕ and η will denote the scalar fields (or components of scalar fields) that cause the successive breakdowns and H and K will denote the little algebras of the constant vectors $\overset{\circ}{\phi}$ and $\overset{\circ}{\eta}$. Then $H \cap K$ is the final little algebra, and if $\perp$ denotes orthogonal complement in the Lie algebra of G, one sees that the intermediate and final Goldstone directions are

$$H^{\perp}\overset{\circ}{\phi} \quad \text{and} \quad (H \cap K)^{\perp}(\overset{\circ}{\phi}+\overset{\circ}{\eta}), \qquad (12.34)$$

respectively. Furthermore, since the intermediate ones correspond to the gauge fields that become massive after the first breakdown, one sees that the generators in $H^{\perp}$ have the quantum numbers necessary to mediate baryon decay.

Now since $(H \cap K)^{\perp}$ may be expressed in the form

$$(H \cap K)^{\perp} = H^{\perp}+(H \cap K^{\perp}), \qquad (12.35)$$

and H annihilates $\overset{\circ}{\phi}$ by definition, the final Goldstone directions may be divided into two sets

$$H^{\perp}(\overset{\circ}{\phi}+\overset{\circ}{\eta}) \quad \text{and} \quad (H \cap K^{\perp})\overset{\circ}{\eta}, \qquad (12.36)$$

where the first set (which may be regarded as rotations of the intermediate Goldstone directions) carry baryon-decay charges, and the second set do not. These two sets are just the sets $(\theta, \rho(\alpha))$ and $(\rho(\beta), \nu)$ respectively in the model of the previous section

The directions $\overset{\circ}{\phi}$ and $\overset{\circ}{\eta}$ do not themselves carry baryon-decay charges, but because the $H^{\perp}$ do, the directions $H^{\perp}\overset{\circ}{\phi}$ and $H^{\perp}\overset{\circ}{\eta}$ carry them. Among these, the combinations $H^{\perp}(\overset{\circ}{\phi}+\overset{\circ}{\eta})$ are Goldstone directions and thus, from the Higgs mechanism, they correspond to gauge rather than scalar fields (indeed to the gauge fields that finally mediate baryon decay). Only the scalar fields orthogonal to these Goldstone fields in the space spanned by $H^{\perp}\overset{\circ}{\phi}$ and $H^{\perp}\overset{\circ}{\eta}$ are true scalar fields which mediate baryon decay. These are the fields $\rho_{\perp}(\alpha)$ in the model of the previous section, and one may think of them as the Goldstone fields which become massive after the second breakdown. It is not difficult to see that there are just dim $(H^{\perp}\overset{\circ}{\eta})$ such fields and the important point for the gauge hierarchy problem (next section) is that there is always at least one of them. Otherwise one would have $H^\perp \mathring{\eta} = 0$ for all $H^\perp$ and since, from the lemma of section 8.6 the closure under commutation of $H^\perp$ is G itself, this would imply $G\mathring{\eta} = 0$, which would contradict the assumption that $\mathring{\eta}$ causes a symmetry breakdown. On the other hand, the masses acquired by the intermediate Goldstone fields after the second breakdown must be proportional to the cross-couplings which reduce the $G(\phi) \times G(\eta)$ symmetry for the fields ϕ and η separately to a common G-symmetry, because otherwise these fields would be Goldstone fields for $G(\phi)$-symmetry alone. This result has already been seen in the model of the previous section where the fields $\rho_\perp(\alpha)$ have masses proportional to the coupling constant ω.

12.5 The gauge hierarchy problem

In grand unified theory the spontaneous symmetry breakdown is assumed to occur in successive stages e.g. $SO(10) \to SU(5) \to S(U(3) \times U(2)) \to U(3)$ and the set of successive symmetry groups $SO(10)$, $SU(5)\dots$ is called a gauge hierarchy. Each stage in the hierarchy is defined by the *energy scale*, which is the vacuum value $|\mathring{\phi}|$ of the scalar field that causes it. (Recall that scalar fields have the dimensions of mass.) For example, in the $SU(5)$ model the first breakdown $SU(5) \to S(U(3) \times U(2))$ occurs for $|\mathring{\phi}| \sim 10^{15}$ GeV and the second breakdown $S(U(3) \times U(2)) \to U(3)$ for $|\mathring{\phi}| \sim 10^2$ GeV. If the successive energy scales correspond to decreasing temperatures, as happens in superconductivity, and is supposed to happen in cosmology, then the stages may be thought of as successive stages in time, as well as energy.

The problem with GUTs is that the ratio of the successive energy scales is enormous ($\sim 10^{13}$) and in order to produce such a ratio the parameters in the potential must be fine tuned to order 10^{-26}, as will now be discussed. Let G be a simple group and

$$V = V_4(\phi) + V_4(\eta) + V_4(\phi, \eta) + m(\phi, \phi) + \mu(\eta, \eta), \qquad (12.37)$$

a G-invariant potential which causes a two-stage breakdown of G through the fields ϕ and η (and which, for simplicity, is assumed to have no cubic terms). As a specific example one may keep in mind the $SU(n)$ defining + adjoint representation model of the previous section with the sequence $SU(n) \to S(U(p) \times U(q+1)) \to S(U(p) \times U(q))$ $(|\mathring{\phi}| > |\mathring{\eta}|)$, which reduces to $SU(5) \to S(U(3) \times U(2)) \to U(3)$ for $n = 5, p = 3$. Now if one is only interested in the scales (s and t say) of ϕ and η then V may be written in the form

$$V = A(\mathring{\phi}) s^4 + 2B(\mathring{\phi}, \mathring{\eta}) s^2 t^2 + C(\mathring{\eta}) t^4 + ms^2 + \mu t^2, \qquad (12.38)$$

where $\hat{\phi}$ and $\hat{\eta}$ are the normalized fields $\hat{\phi} = \phi/s$, $\hat{\eta} = \eta/t$, and A, B, C depend only on these and dimensionless coupling constants. The extremal equations for s and t are evidently

$$2As^2 + 2Bt^2 + m = 2Ct^2 + 2Bs^2 + \mu = 0, \qquad (12.39)$$

from which one obtains the ratio

$$\frac{t^2}{s^2} = \frac{A\mu - Bm}{Cm - B\mu}. \qquad (12.40)$$

Thus, if the ratio of the scales is to be of order 10^{-13}, the parameters must be fine tuned so that the quantity on the right-hand side of (12.40) is of order 10^{-26}. A natural way to arrange this would be to let the interaction term B be equal to zero (at least to order 10^{-26}) and the mass parameters μ, m be in the ratio 10^{-26}. However, this is not possible, because, for two separate reasons, B must be at least of order e^2, where e is the gauge-coupling constant. First, the masses of the gauge and scalar fields that mediate proton decay are of order e^2s^2 and Bs^2 respectively (since, as discussed in the previous section, the scalars would be Goldstone fields if B were zero), and hence if the scalar-mediated decay is to be as slow as the gauge-field-mediated decay, $B \gtrsim e^2$. Second, even if B is chosen to be of order 10^{-26} at the classical level, the radiative corrections with the gauge field will, in general, produce a coupling of order $B \approx e^2$. More generally, for any choice of A, B, C, m, μ that produces a ratio s^2/t^2 in (12.40) of order 10^{-26} at the classical level, the radiative corrections will, in general, change and the ratio to order $e^2 \approx 10^{-2}$ at the one-loop level and will thus destroy the fine tuning.

The problem is actually two-fold. First, even at the tree level there must be a very fine adjustment of parameters to avoid rapid proton decay, and there is (so far) no *a priori* principle to produce such an adjustment, e.g. to make $A\mu = mB$. Second, even if the adjustment is made at the tree-level, it will not, in general, be preserved by the radiative corrections. For example, unless the renormalization group equation for t^2/s^2 is

$$\mu \frac{\mathrm{d}}{\mathrm{d}\mu}(t^2/s^2) = 0 \qquad (12.41)$$

(up to thirteen loops!) a small variation in the scale parameter μ will destroy the adjustment. Of course, in principle, appropriate subtractions could be made at each order in the loop expansion, but that would be too *ad hoc* for comfort. One proposal to solve the radiative problem is to make the system supersymmetric (Witten, 1981) because supersymmetry has the property that radiative corrections do not change the parameter ratios. However, this entails all the problems of supersymmetry (how to break it, why is it not observed, etc.). Furthermore, for ordinary (non-gravitational) supersymmetry it only solves the problem of the *stability* of the fine tuning with respect to radiative corrections, but does not explain why the tuning has to be so fine in the first place. More recently attempts have been made to motivate the fine tuning by using supergravity (Ross, 1984; Hall, Lykken and Weinberg, 1983), but these attempts are still at a tentative stage. It is probably fair to say that, at the present time, the gauge-hierarchy problem remains unsolved, and could be serious enough to undermine the whole grand unification program.

Exercise

12.1. Let

$$V(\phi) = \frac{h}{4!}(\mathrm{tr}\,\phi^2)^2 + \frac{f}{3!}(\mathrm{tr}\,\phi^3) + \frac{m}{2}(\mathrm{tr}\,\phi^2), \quad h > 0, f > 0,$$

be the potential for a field ϕ in the adjoint representation of $SU(3)$,

$$\phi(x) = \Sigma\,\phi_\alpha(x)\,\sigma_\alpha, \quad \alpha = 1, \dots, 8.$$

Show that the potential minimum is at $\overset{\circ}{\phi} = 0$ for $m > f^2/18h$ and at $\overset{\circ}{\phi} = [f + (f^2 - 16mh)^{\frac{1}{2}}]\,y/4h$, where $y = \mathrm{diag}(1, 1, -2)$ for $m < f^2/18h$.

12.2. Identify the Goldstone fields and the polar field π in the second case of exercise 12.1. Show that there is one other mass multiplet and compute the mass of this multiplet and of π.

12.3. Show that the symmetry-breaking patterns of the potential

$$V = \frac{h}{4}(\mathrm{tr}\,\phi^2)^2 + \frac{f}{4!}(\mathrm{tr}\,\phi^4) \mp \frac{m}{2}(\mathrm{tr}\,\phi^2)$$

for the symmetric and antisymmetric tensor representations of $SU(n)$ are

Symmetric tensor: $SU(n) \to O(n), f > 0, SU(n) \to SU(n-1), f < 0$,
Antisymmetric tensor: $SU(n) \to Sp(2l), f > 0, SU(n) \to SU(n-2), f < 0$.
(Li, 1974)

12.4. Let

$$V = \frac{h}{4!}(\phi, \phi)^2 + \frac{m}{2}(\phi, \phi) + \frac{g}{2}(\phi, \eta)^2 + \frac{\lambda}{4!}(\eta, \eta)^2 + \frac{\mu}{2}(\eta, \eta),$$

be the potential for fields ϕ and η in the fundamental representation of $SU(n)$. Given that m and μ are negative show that the little group of the minimum is $SU(n-1)$ for $g < 0$, and $SU(n-2)$ for $g > 0$, and identify the Goldstone and polar fields in each case.

References

Abud, M., Anastaze, G., Eckert, P. and Ruegg, H. (1984) *Phys. Letts.*, **142B**, 371.
Abud, M. and Sartori, G. (1983) *Ann. Phys.*, **150**, 307–72.
Adler, S. (1969) *Phys. Rev.* **177**, 2426.
Adler, S. (1970) In *Lectures on Elementary Particles and Quantum Field Theory* (ed. S. Deser, M. Grisaru and H. Pendleton), MIT Press, Cambridge, Mass.
Amit, D. (1978) *Field Theory, Renormalization Group and Critical Phenomena*, McGraw-Hill, New York.
Antoine, J-P. and Speiser, D. (1964) *J. Math. Phys.*, **5**, 1560.
Appelquist, T. and Carrazone, J. (1975) *Physical Review*, **D11**, 2856.
Arnison, G. *et al.* (UA1) (1983) *Phys. Lett.*, **122B**, 103 and **129B**, 273.
Baaklini, N. (1980) *Phys. Rev.* **D21**, 1932.
Bacry, H. (1977) *Lectures on Group Theory and Particle Theory*, Gordon and Breach, New York.
Bagger, J. and Dimopoulos, S. (1984) *Nucl. Phys.* **B244**, 247.
Bagger, J. and Wess, J. (1983) *Supersymmetry and Supergravity*, Princeton Series in Physics, Princeton University Press, Princeton.
Bagnera, P. and Banner, M. *et al.* (UA2) (1983) *Phys. Lett.*, **122B**, 476 and **129B**, 130.
Balachandran, A. *et al.* (1984) **29D** *Physical Review*, 2919, 2936.
Bander, M. (1981) *Physics Reports*, **75**, 4.
Bardeen, W. (1973) In *Proc. XVIth Internal Conference on High Energy Physics, Chicago–Batavia* (ed. J. Jackson and A. Roberts) National Accelerator Laboratory Publication, 111. *Phys. Rev.* **184**, 1848.
Barut, A. and Raczka, R. (1965) *Proc. Roy. Soc.* **A287**, 519.
Barut, A. and Raczka, R. (1977) *Theory of Group Representations and Applications*, Polish Scientific Publishers, Warsaw.
Bassetto, A., Ciafolini, M. and Marchesini, G. (1983) *Phys. Reports*, **100**, 4.
Becher, P., Bohm, M. and Joos, H. (1984) *Gauge Theories of the Strong and Weak Interactions*, Wiley, New York.
Bell, J. and Jackiw, R. (1969) *Nuovo Cim.*, **60A**, 47.
Biedenharn, L. (1963) *J. Math. Phys.* **4**, 436.
Bilenky, S. and Hösek, J. (1982) *Physics Reports*, **90**, 74.
Bincer, A. and Schmidt, J. (1984) *J. Math. Phys.*, **25**, 2367.
Boerner, H. (1970) *Representations of Groups*, North-Holland, Amsterdam.
Bogolibov, N. and Shirkov, D. (1954) *Introduction to the Theory of Quantized Fields*, Interscience, New York.
Bott, R. (1982) *Bull. Am. Math. Soc.*, **7**, 331.
Buccella, F., Cocco, L. and Wetterich, C. (1984) *Nucl. Phys.*, **B243**, 273.
Buccella, F., Ruegg, H. and Savoy, C. A. (1980) *Nucl. Phys.*, **B169**, *Phys. Lett.* **94B**, 491.
Buras, A. (1980) *Rev. Mod. Phys.*, **52** (1) 199.

References   155

Buras, A., Ellis, J., Gaillard, M. and Nanopoulos, D. (1978) *Nucl. Phys.*, **B135**, 66.
Burzlaff, J., Murphy, T. and O'Raifeartaigh, L. (1985) *Phys. Lett.*, **154B**, 159.
Cabibbo, N. (1963) *Phys. Rev. Lett.*, **10**, 531.
Cahn, R. (1984) *Semi-Simple Lie Algebras*, Benjamin, New York (Frontiers in Physics Series, Vol. 59).
Callan, C. (1982) *Phys. Rev.*, **25D**, 2141, **26D**, 2058.
Callan, C., Dashen, R. and Gross, D. (1976) *Phys. Lett.*, **63B**, 334.
Carruthers, P. (1971) *Spin and Isospin in Particle Physics*, Gordon and Breach, New York.
Chaichan, M., Kolmakov, Y. and Nelipa, N. (1982) *Nucl. Phys.*, **B202**, 126.
Chau, L-L. (1983) *Physics Reports*, **95** (1) 1.
Close, F. (1979) *Introduction to Quarks and Partons*, Academic Press, New York.
Creutz, M. (1983) *Quarks, Gluons and Lattices*, Cambridge University Press.
Daniel, M. and Viallet, C. (1980) *Rev. Mod. Phys.*, **52** (1) 175.
Davidson, A., Nair, V. and Wali, K. C. (1984) *Phys. Rev.*, **D29**, 1505.
Dine, H., Fischler, W. and Srednicki, M. (1981) *Phys. Lett.*, **104B**, 199.
Dolgov, A. and Zeldovich, Y. (1981) *Rev. Mod. Phys.*, **53** (1) 1.
Dynkin, E. (1975) *Amer. Math. Soc. Trans. Ser.*, **2** (6) 111, 245.
Ecker, G. (1984) *Acta Physica Polonica*, **B15**, 179.
Ellis, J. (1981) In *Gauge Theories and Experiments at High Energies*, *XXI Scottish Universities Summer School* (ed. K. Bowler and D. Sutherland), SUSS Publication, Edinburgh.
Fayet, P. and Ferrarra, S. (1977) *Physics Reports*, **32C**, 249.
Ferrarra, S. (1984) *Supersymmetry*, World Scientific, Singapore.
Fetter, A. and Walecka, J. (1971) *Quantum Theory of Many Particle Systems*, McGraw-Hill, New York.
Fishbane, P. and Meshkov, S. (1984) Comments on Nuclear and Particle Physics **13** (6) 285.
Frampton, P. (1979) *Phys. Lett.*, **88B**, 299.
Fritzsch, H., Gell-Mann, M. and Leutwyler, H. (1973) *Phys. Lett.*, **47B**, 365.
Fritzsch, H. and Minkowski, P. (1975) *Ann. Phys.*, **93**, 193.
Fritzsch, H. and Minkowski, P. (1981) *Physics Reports*, **73** (2) 67.
Fröhlich, J. (1982) *Nucl. Phys.*, **B200**, 281.
Fujimoto, Y. (1981) *Nucl. Phys.*, **B182**, 242.
Gasiorowicz, S. and Geffen, D. (1969) *Rev. Mod. Phys.*, **41**, 531 (p. 536).
Gasser, J. and Leutwyler, H. (1982) *Physics Reports*, **87**, 3.
Gell-Mann, M. and Glashow, S. (1961) *Ann. Phys.*, **15**, 437.
Gell-Mann, M. and Ne'eman, Y. (1964) *The Eightfold Way*, Benjamin, New York.
Gell-Mann, M., Ramond, P. and Slansky, R. (1978) *Rev. Mod. Phys.*, **50**, 721.
Georgi, H. (1979) *Nucl. Phys.*, **B156**, 126.
Georgi, H. and Glashow, S. (1972) *Phys. Rev.*, **D6**, 429.
Georgi, H. and Glashow, S. (1974) *Phys. Rev. Lett.*, **32**, 438.
Georgi, H., Quinn, H. and Weinberg, S. (1974) *Phys. Rev. Lett.*, **33**, 451.
Gilmore, R. (1974) *Lie Groups, Lie Algebras*, Wiley, New York.
Glashow, S. (1961) *Nucl. Phys.*, **22**, 579.
Glashow, S. (1980) *Rev. Mod. Phys.*, **52** (3) 539.
Glashow, S., Iliopolous, J. and Maiani, L. (1970) *Phys. Rev.*, **D2**, 1285.
Goddard, P. and Olive, D. (1978) *Rep. Prog. Phys.*, **41**, 1357.

Goldstone, J., Salam, A. and Weinberg, S. (1962) *Phys. Rev.*, **127**, 965.
Gourdin, M. (1982) *Basic Lie Groups*, Editions Frontiéres, Gif-sur-Yvette.
Green, M. and Schwarz, J. (1984) *Phys. Lett.*, **149B**, 117.
Greenberg, O. (1978) *Ann. Rev. Nucl. Sci.*, **28**, 327.
Gribov, V. (1977) *Instability of Non-Abelian Gauge Theories and Impossibility of Choice of Coulomb Gauge*, Stanford Linear Accelerator Translation 176, Stanford, California.
Gross, D. and Jackiw, R. (1972) *Phys. Rev.*, **D6**, 477.
Gross, D. and Wilczek, F. (1973) *Phys., Rev.*, **D8**, 3633.
Gursey, F. and Sikivie, P. (1976) *Phys. Rev. Lett.*, **36**, 775.
Gursey, F. and Sikivie, P. (1977) *Phys. Rev.*, **D16**, 816.
Hall, L., Lykken, J. and Weinberg, S. (1983) *Phys. Rev.*, **27D**, 2359.
Helgason, S. (1978) *Differential Geometry, Lie Groups and Symmetric Spaces*, Academic Press, New York.
Houston, P. and Sen, S. (1984) *Journal of Physics*, **A17**, 1163.
Huang, K. (1982) *Quarks, Leptons and Gauge-Fields*, World Scientific, Singapore.
Humphreys, J. (1972) *Introduction to Lie Algebras and Representation Theory*, Springer, New York.
Hung, P. and Sakurai, J. (1981) *Ann. Rev. Nucl. Sci.*, **31**, 375–438.
Isgur, N. and Karl, G. (1983) *Physics Today*, **36** (11) 36–43.
Kaul, R. (1983) *Rev. Mod. Phys.*, **55**, 449.
Kim, C. and Roiesnel, C. (1980) *Phys. Lett.*, **93B**, 343.
Kim, J. E., Langacker, P., Levine, M. and Williams, H. (1981) *Rev. Mod. Phys.*, **53** (2) 211.
Kim, J. S. (1982) *Nucl. Phys.*, **B196**, 285, **B207**, 374.
Kobayashi, M. and Maskawa, T. (1973) *Prog. Theor. Phys.*, **49**, 652.
Koca, M. (1981) *Phys. Rev.*, **D24**, 2645.
Kogut, J. (1983) *Ann. Rev. Nucl. Sci.*, **55** (3) 775.
Kolb, E. and Turner, M. (1983) *Ann. Rev. Nucl. Sci.*, **33**, 645.
Kostant, B. (1959) *Trans. Amer. Math. Soc.*, **93**, 53.
Langacker, P. (1981) *Physics Reports*, **72** (4) 185.
Lee, T. D. and Yang, C. N. (1956) *Nuovo Cim.*, **10** (3) 749.
Leibrandt, G. (1975) *Rev. Mod. Phys.*, **47** (4) 849.
Li, L. F. (1974) *Phys. Rev.*, **9D**, 1723.
Lichtenberg, D. (1970) *Unitary Symmetry and Elementary Particles*, Academic Press, New York.
MacKay, M. and Patera, J. (1981) *Tables of Dimensions, Indices, Branching Rules for Representations of Simple Algebras*, Dekker, New York.
Marciano, W. and Sirlin, A. (1984) *Phys. Rev.*, **29D**, 945.
Marshak, R., Riazud-din and Ryan, C. (1969) *Theory of Weak Interactions*, Wiley, New York.
Michel, L. (1953) *Nuovo Cim.*, **10**, 319.
Michel, L. (1971) *C.R. Acad. Sci. Paris*, **272**, 433.
Michel, L. (1972) *Non-Linear Group Action*, Universities Press, Jerusalem.
Michel, L. (1977) In *Group Theoretical Methods in Physics*, Academic Press, New York.
Michel, L. (1980) Proc. of Colloq. in Honour of A. Visconti, CNRS, Marseille.
Michel, L. and Radicati, L. (1971) *Ann. Phys.*, **66**, 758.
Michel, L. and Radicati, L. (1973) *Ann. Inst. Henri Poincaré*, **18**, 185.
Mithra, P. (1984) Univ. of Halifax, Nova Scotia, Private Communication.

Mohapatra, R. (1985) in Flavour-Mixing in Weak Interactions (ed. L-L. Chau, Plenum, New York).
Montgomery, D. and Zippen, L. (1955) *Topological Transformation Groups*, Interscience, New York.
Mostow, G. (1958) *Amer. J. Math.*, **80**, 331.
Mueller, A. (1981) *Physics Reports*, **73**, 4.
Murphy, T. and O'Raifeartaigh, L. (1983), *Nucl. Phys.*, **B229**, 509; with Yamada, M. (1984), *Nucl. Phys.* **B248**, 356.
Nash, C. and Sen, S. (1983) *Topology and Geometry for Physicists*, Academic Press, New York.
Okubo, S. (1962) *Prog. Theor. Phys.*, **27**, 949.
Olive, D. (1982) In *Monopoles in Quantum Field Theory*, World Scientific, Singapore, p. 184.
Pati, J. (1978) In *Topics in Quantum Field and Gauge Theories*, Springer Lecture Notes, Berlin, p. 221.
Pati, J. and Salam, A. (1973) *Phys. Rev.*, **D8**, 1240.
Peccei, R. and Quinn, H. (1979) *Phys. Rev. Lett.*, **38**, 1440.
Politzer, D. (1974) *Physics Reports*, **14C**, 129.
Pontryagin, L. (1966) *Topological Groups*, Gordon and Breach, New York.
Primakoff, H. and Rosen, S. P. (1981) *Ann. Rev. Nucl. Sci.*, **31**, 145.
Quinn, H. and Peccei, R. (1979) *Phys. Rev. Lett.*, **38**, 1440.
Racah, G. (1950) *Rendiconti dei Lincei*, **8**, 108.
Racah, G. (1965) *Ergebnisse der Exakten Wissenschaften*, **37**, 28.
Rajpoot, S. (1979) *Phys. Rev.*, **D20**, 1688.
Ramond, P. (1983) *Ann. Rev. Nucl. Sci.*, **33**, 31.
Rebbi, C. (1983) *Lattice Gauge Theories and Monte Carlo Simulations*, World Scientific, Singapore.
Reya, E. (1981) *Physics Reports*, **69**, 3.
Ross, G. (1984) In *Supersymmetry, Supergravity and Related Topics* (XVth GIFT Seminar, Girona, Spain) World Scientific, Singapore.
Rubakov, V. (1981*a*) *Pis'ma, Zh. Eksp. Teor. Fiz.*, **33**, 658.
Rubakov, V. (1981*b*) *JETP Lett.*, **33**, 644.
Rubakov, V. (1982) *Nucl. Phys.*, **B203**, 311.
Ruegg, H. (1980) *Phys. Rev.*, **22D**, 2040.
Salam, A. (1968) In *Nobel Symposium on Elementary Particle Theory*, Wiley Interscience, New York.
Salam, A. (1980) *Rev. Mod. Phys.*, **52**, 525.
Salam, A. and Ward, J. (1964) *Phys. Lett.*, **13**, 168.
Samios, N., Goldberg, M. and Meadows, B. (1974) *Rev. Mod. Phys.*, **46**, 49.
Schwarz, G. (1975) *Topology*, **14**, 63.
Siebenthal, J. de (1956) *Comment. Math. Helv.*, **31**, 41.
Singer, I. (1978) *Comm. Math. Phys.*, **60**, 7.
Slansky, R. (1981) *Physics Reports*, **79**, No. 1.
Smale, S. (1977) *Bull. Am. Math. Soc.*, **83**, No. 4.
Söding, P. and Wolf, G. (1981) *Ann. Rev. Nucl. Sci.*, **31**, 231.
Stech, B. (1980) In *Unification of the Fundamental Particle Interactions*, Plenum, New York, p. 23.
Steigman, G. (1979) *Ann. Rev. Nucl. Sci.*, **29**, 313.
Sternberg, S. (1983) *Lectures in Differential Geometry*, Chelsea, New York, p. 232.
Symanzik, K. (1969) In *Local Quantum Theory* (ed. R. Jost) Academic Press, New York.

Symanzik, K. (1973) *Commun. Math. Phys.*, **34**, 7.

Tits, J. (1967) *Tabellen zu den Einfachen Gruppen und ihren Darstellungen*, Lecture Notes in Mathematics, No. 40, Springer, Berlin.

Trilling, G. (1981) *Physics Reports*, **75**, 2.

Tuzzi, T. (1985) University of Naples Doctoral Thesis.

Utiyama, R. (1956) *Phys. Rev.*, **101**, 1597.

Varadarajan, V. (1974) *Lie Groups, Lie Algebras and their Representations*, Prentice Hall, New Jersey.

Wawrzynczyk, A. (1984) *Group Representations and Special Functions*, Reidel, Dordrecht.

Weil, A. (1953) *l'Integration dans les Groups Topologiques*, Hermann, Paris.

Weinberg, E. (1980) *Nucl. Phys.*, **B167**, 500.

Weinberg, S. (1967) *Phys. Rev. Lett.*, **19**, 1264.

Weinberg, S. (1973) *Phys. Rev.*, **D8**, 3497.

Weisberger, W. (1981) *Phys. Rev.*, **24**, 481, 1617.

Weyl, H. (1929) *Z. Phys.*, **56**, 330.

Wilczek, F. (1982) *Ann. Rev. Nucl. Sci.*, **32**, 172.

Wilczek, F. and Zee, A. (1982) *Phys. Rev.*, **D25** (2) 553.

Witten, E. (1981) *Nucl. Phys.*, **B188**, 513.

Witten, E. (1982) *Morse Theory and Supersymmetry, J. Diff. Geom.* **17**, 661.

Wohl, C. (1984) Review of Particle Properties, *Rev. Mod. Phys.*, 56 (2) Part II.

Wybourne, B. (1974) *Classical Groups for Physicists*, Wiley-Interscience, New York.

Yang, C. N. and Mills, R. (1954) *Phys. Rev.*, **96**, 191.

Zumino, B. (1984) *Chiral Anomalies and Differential Geometry in Relativity Groups and Topology*, North-Holland.

Suggestions for further reading

GROUP THEORY

Hausner, M. and Schwartz, J. (1968) *Lie Groups, Lie Algebras*, Gordon and Breach, New York.
Hermann, R. (1966) *Lie Groups for Physicists*, Benjamin, New York.
Jacobson, N. (1962) *Lie Algebras*, Wiley-Interscience, New York.
Murnaghan, F. (1938) *Theory of Group Representations*, Johns Hopkins University Press, Baltimore.
Naimark, M. and Stern, A. (1982) *Theory of Group Representations*, Springer, New York.
Wan, Z-X. (1975) *Lie Algebras*, Pergamon Press, New York.
Weyl, H. (1946) *Classical Groups*, Princeton University Press, Princeton, New Jersey.
Zassenhaus, H. (1949) *Theory of Groups*, Chelsea, New York.
Zelebenko, D. (1973) *Compact Lie Groups and their Representations*, American Math. Soc., Providence, R.I.

GENERAL GAUGE THEORY

Abers, E. and Lee, B. (1973) *Physics Reports*, 9, 1.
Aitcheson, I. and Hey, A. (1982) *Gauge Theory in Particle Physics*, Hilger, Bristol.
Bernstein, J. (1974) Spontaneous symmetry breaking, gauge theories, Higgs mechanism and all that, *Rev. Mod. Phys.*, 46 (1) 7–48.
Collins, J. (1984) *Renormalization*, Cambridge University Press.
Faddeev, L. and Slavnov, A. (1980) *Gauge Fields, Introduction to Quantized Theory*, Benjamin/Cummings, New York.
Itzykson, C. and Zuber, J. B. *Quantum Field Theory*, McGraw-Hill, New York.
Jackiw, R. (1980) Introduction to Yang–Mills quantum theory, *Rev. Mod. Phys.*, 52 (4) 661–73.
Konopleva, N. and Popov, V. (1981) *Gauge Fields*, Harwood, New York.
Leites-Lopez, J. (1981) *Gauge Field Theories, An Introduction*, Pergamon, New York.
Moriyasu, K. (1983) *An Elementary Primer of Gauge Theory*, World Scientific, Singapore.

GAUGE THEORY FOR CLASSICAL FIELDS

Actor, A. (1979) Classical solutions of Yang–Mills theories, *Rev. Mod. Phys.*, 51 (3) 461–526.
Carmelli, M. (1982) *Classical Fields*, Wiley-Interscience, New York.
Chaohao, G. (1981) Classical Yang–Mills fields, *Physics Reports*, 80 (4) 251.

ELECTROWEAK INTERACTIONS

Bailin, D. (1982) *Weak Interactions*, Hilger, Bristol.
Beg, M. and Sirlin, A. (1982) Gauge theories of weak interactions, *Physics Reports*, **88** (1) 1.
Commins, E. (1973) *Weak Interactions*, McGraw-Hill, New York.
Lai, C. (1981) *Gauge Theory of the Weak and Electromagnetic Interaction*, World Scientific, Singapore.
Taylor, J. C. (1976) *Gauge Theories of Weak Interactions*, Cambridge University Press.
Weinberg, S. (1980) Conceptual foundations of the unified theory of weak and electromagnetic interactions, *Rev. Mod. Phys.*, **52** (3) 515–24.

STRONG, WEAK AND ELECTROMAGNETIC INTERACTIONS

Annual Workshops on Grand Unification. Latest is the *Vth Workshop at Brown University 1984* (ed. K. Kang, H. Fried and P. Frampton) World Scientific, Singapore.
Becher, P., Böhm, M. and Joos, H. (1981) *Eichtheorien der Starken und Electroschwachen Wechselwirkungen*, Teubner, Stuttgart; (1984) *Gauge Theories of Strong and Electroweak Interactions*, Wiley, New York.
Cheng, T-P. and Li, L-F. (1984) *Gauge Theory of Elementary Particle Physics*, Clarendon Press, Oxford.
Farhi, E. and Susskind, L. (1981) Technicolour, *Physics Reports*, **74** (3) 277.
Ferrara, S., Ellis, J. and van Nieuwenhuizen, P. (eds.) (1980) *Unification of the Fundamental Particle Interactions*, Plenum, New York.
Georgi, H. (1983) *Lie Algebras in Particle Physics*, Benjamin/Cummings, New York.
Halzen, F. and Martin, A. (1984) *Quarks and Leptons*, Wiley, New York.
Lai, C. and Mohopatra, R. (1981) *Gauge Theories of the Fundamental Interactions*, World Scientific, Singapore.
Quigg, C. (1983) *Gauge Theories of Strong, Weak and Electromagnetic Interactions*, Benjamin/Cummings, New York.
Zee, A. (ed.) (1982) *Unity of Forces in the Universe*, Vols I and II, World Scientific, Singapore.

COSMOLOGY

Dolgov, A. and Zeldovich, Ya. (1981) *Rev. Mod. Phys.*, **53**, 1–41.
Gibbons, G., Hawking, S. and Siklos, S. (1983) *The Very Early Universe*, Cambridge University Press.
Harrison, E. (1981) *Cosmology*, Cambridge University Press.
Hawking, S. and Ellis, G. (1973) *Large-Scale Structure of Space–Time*, Cambridge University Press.
Novikov, I. and Zeldovich, Ya. (1983) *The Structure and Evolution of the Universe*, University of Chicago Press.
Peebles, J. (1980) *The Large Scale Structure of the Universe*, Princeton University Press, New Jersey.
Schramm, D. (1983) The early Universe and high energy physics, *Rev. Mod. Phys.*, **36** (4) 27–33.
Steigman, G. (1979) Cosmology Confronts Particle Physics, *Ann. Rev. Nucl. Sci.*, **29**, 313–38.
Weinberg, S. (1972) *Gravitation and Cosmology*, Wiley, New York.

GRAVITATION AND GAUGE THEORY

Hehl, F., von der Heyde, P., Kerlick, G. and Nester, J. (1976) *Rev. Mod. Phys.*, **48**, 393–416.

Isham, C., Penrose, R. and Sciama, D. (1981) *Quantum Gravity*, Oxford University Press.

Ivanenko, D. and Sardanashvily, G. (1983) Gauge treatment of gravity, *Physics Reports*, **94** (1) 1.

Penrose, R. and Rindler, W. (1984) *Spinors and Space–Time*, Volume 1, *Two-spinor calculus and relativistic fields*, Cambridge University Press.

Glossary

162

Compact Lie algebra: Lie algebra of a compact group.

Idempotent (in symmetric algebra): Elements of symmetric algebra whose squares are proportional to themselves.

Invariant subalgebras: Subalgebras of Lie algebras which are invariant (as a set) with respect to commutation by the whole algebra.

Jacobi identity: Cyclic, bilinear identity for the structure constants of a Lie algebra.

Primitive charges: Base elements for Cartan subalgebra which are not orthogonal but are normalized Cartan bases for the primitive $SU(2)$ subalgebras.

Primitive roots: l positive roots such that all the positive roots of a compact simple Lie algebra can be expressed as linear combinations of them with positive integer coefficients, where l is the rank.

Primitive $SU(2)$ subalgebra: Subalgebra formed by the element corresponding to a primitive root, its conjugate, and the Cartan element formed by the commutator of these two.

Roots: Simultaneous eigenvalues of the elements of the Cartan subalgebra of a compact simple Lie algebra in the adjoint representation.

Root diagram: Diagrammatic representation of roots as vectors in an l-dimensional Euclidean space, where l is the rank.

Semi-simple Lie algebra: One for which the Cartan metric is not degenerate, or, equivalently, one which admits no abelian invariant subalgebra.

Structure constants: Set of constants (satisfying the Jacobi identity) which define the multiplication law of a Lie group in the neighbourhood of the identity.

Weights: Simultaneous eigenvalues of the elements of the Cartan subalgebra in any hermitian irreducible representation of a compact Lie algebra.

Primitive weights: Weights which are dual to the primitive charges in Lie algebra (they are the highest weights for the primitive representations).

Highest weights: Weights which are the highest in a given irreducible representation (with respect to a given ordering of the basis of the Cartan subalgebra).

FIELDS (PARTICLES)

Baryons: Strongly interacting observable fermions.

Bosons: Integer-spin fields.

Chiral fields: Spin-$\frac{1}{2}$ fields which have definite chirality but not definite parity, i.e. which are eigenstates of the Dirac operator γ_5 and thus belong to either the $D(\frac{1}{2}, 0)$ or $D(0, \frac{1}{2})$ representation of Lorentz group (but not both).

Dirac fields: Spin-$\frac{1}{2}$ fields of definite parity belonging to the $D(\frac{1}{2}, 0) + D(0, \frac{1}{2})$ representation of Lorentz group.

Fermions: Half-odd-integer spin fields.

Gauge potentials: Vector fields A which have the non-covariant transformation law $A_\mu \to U A_\mu U^{-1} + U \partial_\mu U^{-1}$ under gauge transformations.

Gauge fields: Antisymmetric tensor fields $F_{\mu\nu}$, constructed from the gauge potentials and having the covariant transformation law $F_{\mu\nu} \to U F_{\mu\nu} U^{-1}$ under gauge transformations.

Generations: Associated pairs of leptons and coloured quarks e.g. $\{(e, \nu)(u^a, d^a)\}$, for which a renormalizable (anomaly-free) $S(U(3) \times U(2))$ gauge theory can be constructed.

Glueballs: Composite states that are made out of gluons and are colour-neutral.

Gluons: Gauge fields associated with colour symmetry.
Goldstone fields: Massless scalar fields which must appear when a continuous
    symmetry is spontaneously broken.
Hadrons: Strongly interacting observable fields of any spin.
Higgs fields: Scalar fields which are introduced in order to generate a
    spontaneous symmetry breakdown.
Leptons: Elementary fields which do not interact strongly, i.e. are neutral with
    respect to colour symmetry (effectively the electron, muon, tauon and their
    neutrinos).
Matter fields: All fields except gauge fields (effectively spin-$\frac{1}{2}$ and scalar fields).
Mesons: Strongly interacting observable bosons.
Monopoles: Composite fields corresponding to static, finite energy, topologically
    charged, solutions of the classical field equations. (The name comes from
    the fact that in some cases the topological charge may be identified as a
    magnetic charge.)
Neutrinos: Spin-$\frac{1}{2}$, electrically neutral (and probably massless) partners of the
    electron, muon (and probably tauon) in weak interactions.
Quarks: Spin-$\frac{1}{2}$, strongly interacting (i.e. coloured) fields, out of which all the
    observed strongly interacting fields (hadrons) are supposed to be built.
Radiation fields: Gauge fields.
Weyl fields: Chiral fields.

GROUP TERMINOLOGY

Automorphism: Mapping of group onto itself which preserves the
    multiplication law.
Centre: Maximal closed subgroup which commutes with all elements of group
    (it is, of course, invariant and abelian).
Closed subgroup: Subgroup of topological group which is closed with respect
    to topology of original group.
Conjugation: The transformation $g \to hgh^{-1}$ of one group element by another.
Conjugation class: Orbit of group element with respect to conjugation by whole
    group.
Coset: Orbit of a group element with respect to left or right multiplication by
    all elements of a subgroup.
Group orbit: Set of all points obtained from a given point under any group
    action.
Haar measure: Measure on a topological group which is invariant under left or
    right multiplication (or both).
Homomorphism: Mapping of one group onto another which agrees with the
    multiplication of both groups.
Identity component: Subgroup (invariant) of topological group whose elements
    are connected continuously to the identity element.
Inner automorphism: Automorphism that can be implemented by conjugation
    with elements of the group itself.
Invariant measure: Haar measure.
Invariant subgroup: Subgroup that is mapped onto itself by conjugation with
    whole group.
Isomorphism: Homomorphism which is one–one.
Kernel of homomorphism: Subgroup (invariant) which maps onto the identity
    element of image.
Quotient Group: Group formed by the cosets of an invariant subgroup.
Semi-direct product group: $K \wedge H$: Group with elements (k, h) where $k \in K$,

$h \in H$ and whose multiplication law is $(k, h) \times (k', h') = (kk', hh'(k))$ where the transformations $h \to h(k)$ form an automorphism of H and a homomorphic image of K.

Unitary trick: Conversion of non-compact semi-simple Lie groups into their compact counterparts by multiplying appropriate base elements of the Lie algebra by factors $i = \sqrt{-1}$.

PHYSICS TERMINOLOGY

Cabibbo angle: Angle used to specify the direction of the charged weak current in the adjoint representation of flavour $SU(3)$.

Confinement: Supposed mechanism by which the quarks (and gluons) are permanently bound in colourless composites (hadrons).

Decoupling theorem: Theorem which shows that although the contribution of high-mass (M) fields to low-energy (m) processes in QFT has terms of order $\ln(M/m)$, such terms may be absorbed into the renormalization of the coupling constants leaving only terms of order $m/M \to 0$.

Four-fermi coupling constant G: The coupling constant which appears in the traditional local four-fermion term $G\bar{\psi}\psi\bar{\psi}\psi$ used for the phenomenological description of weak decays at low energies.

Gauge-hierarchy problem: Problem of the huge difference between the energy scales which must be used for the successive symmetry breakdowns in grand unified theories.

Gupta–Bleuler mechanism: Cancellation of longitudinal and timelike components of massless gauge potentials for physical states.

Hessian: Matrix of second derivatives of potential (which becomes the mass matrix at the potential minimum).

Higgs mechanism: Cancellation of massless Goldstone scalars and timelike components of massive gauge potentials for physical states.

Lattice gauge theory: Quantized gauge theory in which Euclidean space is approximated by a (finite) lattice of discrete points.

Noether charges: Charges which generate the symmetries of a Lagrangian and which are conserved by the field equations.

One-loop approximation: Approximation to QFT in which only Feynman diagrams with one closed loop are used (it corresponds to the next order beyond the classical theory in powers of $\hbar$).

Photon–baryon asymmetry: Experimental observation that in the observed universe the ratio photon number/baryon number is of order 10^{10}.

Physical gauge: Gauge in which the cancellation of massless Goldstone scalars and the timelike components of the massive gauge fields is explicit.

Radiative corrections: Corrections to classical theory due to field quantization.

Supersymmetry: Symmetry which relates fields of different spin, notably integer and half-odd-integer spin.

Supergravity: Extension of supersymmetry to include gravity.

Theta (θ)-vacua: Degenerate vacua associated with the existence of topologically non-trivial solutions of YM equations (instantons) in Euclidean four space.

Unitary gauge: Physical gauge.

Weak angle: Angle used to specify the electromagnetic direction in Lie algebra of the electroweak group $U(2)$.

Yukawa coupling: Coupling between spin-$\frac{1}{2}$ and scalar fields, of the general form $G^a_{\alpha\beta}\bar{\psi}_\alpha \psi_\beta \phi_a$ where $G^a_{\alpha\beta}$ are coupling constants.

QUANTUM NUMBERS

$(B-L)$: Baryon minus lepton number, generalized to apply also to bosons.

Charm: Noether charge (internal quantum number) associated with the extension of flavour $SU(3)$ to $SU(4)$.

Chirality: Eigenvalue of Dirac operator γ_5 which describes helicity or 'handedness' (left or right) of a fermion field.

Colour: Noether charge (internal quantum number) associated with the $SU(3)$ 'colour' group i.e. the group whose gauge theory is supposed to describe the strong interactions.

Colour breaking: Breaking of colour symmetry.

CP: C = Charge conjugation operator, P = Parity operator.

Flavours: Generic name for the Noether charges (internal quantum numbers) associated with the 'flavour' symmetry groups (isospin, $SU(3)$ etc.) i.e. those used to label the observed hadrons.

$F\chi$: F = Fermion number, χ = Chirality.

Strangeness: Noether charge (internal quantum number) associated with the extension of isospin $SU(2)$ to flavour $SU(3)$.

REPRESENTATIONS

Adjoint representation: Representation of Lie group on its algebra by conjugation.

Casimir operators: Elements of the enveloping algebra of a Lie algebra which commute with all elements of the Lie algebra.

CUIRs: Continuous unitary irreducible representations (of compact simple Lie groups).

Dominant weights: Weights which are the highest in their Weyl orbits.

Dynkin diagram: Diagrammatic representation of the primitive roots.

Dynkin indices: Indices which label the CUIRs of compact simple groups and indicate how often each primitive representation occurs in its construction.

Fundamental representation (of compact simple Lie group): Defining representation, i.e. lowest dimensional faithful single-valued representation.

Highest weight: Weight in a CUIR which is highest with respect to a fixed ordering of the Cartan algebra.

Index (of a CUIR): Integer which is obtained by suitable renormalization of the Casimir operator.

Levi-Civita symbol: The n-index totally antisymmetric numerical tensor whose components are zero if any two indices are the same and otherwise are ± 1 (according as they are obtained by even or odd permutations of a fixed order).

Linear representation: Representation by matrices or linear operators.

Primitive representations (of compact simple Lie groups): l CUIRs which are constructed by taking antisymmetric products of the fundamental representation, and out of which all the tensor representations can be constructed by taking symmetric products, where l is the rank.

Primitive weights: The highest weights of the primitive CUIRs of compact simple Lie groups. They are dual to the primitive roots.

Pseudo-real representations: Representations which are equivalent to their complex conjugates $R(g)^* = UR(g)\,U^{-1}$, but which cannot be made real because $U^*U = -1$.

Rank-index: Index which indicates how often the fundamental representation of a compact simple Lie group occurs in the construction of a given CUIR (modulo the order of the centre).

Safe groups/CUIRs: Groups/CUIRs for which the third-degree Casimir operator is zero and which is therefore guaranteed not to have any ABJ anomalies.

Spinor representations: Two-valued representations (of the orthogonal groups).

Tensor representations: Single-valued representations of all the compact simple Lie groups (they can be constructed from various products of the fundamental representation).

Triality: Rank index for $SU(3)$.

Vectorlike assignments: Assignments of fermion fields such that the representations of the left- and right-handed fermions are equivalent.

Weights: Simultaneous eigenvalues of the elements of the Cartan algebra in a hermitian representation of a compact simple Lie algebra.

Weight diagram: Diagrammatic representation of weights in an l-dimensional Euclidean space, where $l = $ rank.

SPECIAL GROUPS

Abelian group: Group for which the multiplication law is commutative.

Adjoint group: Quotient group $Q = G/Z$ of group G by its centre Z.

Classical groups: Groups of unitary unimodular, orthogonal, and unitary symplectic $n \times n$ matrices.

Covering group $\tilde{G}$: Simply connected group of which given group G is a homomorphic image.

Exceptional groups: The five simple compact Lie groups G_2, F_4, E_6, E_7, E_8, which are not classical groups.

Gauge group: Lie group whose parameters are space–time dependent.

Global (Lie) group: Lie group whose parameters are not restricted to the neighbourhood of the identity but range over all their values.

Internal (symmetry) groups: Groups which are used to describe physical systems (and their symmetries) but are not associated with transformations of space–time.

Isospin group: An internal $SU(2)$ group used to describe the electromagnetic-charge independence of the strong interactions.

Lie groups: Groups with locally Euclidean topology.

Matrix groups: Groups whose elements are non-singular square matrices.

Normal parameters: Parameters for Lie groups such that each parameter taken alone describes a complete (one-parameter) subgroup.

Order r of Lie group: Minimal number of independent parameters.

Orthogonal groups: Groups whose elements are real orthogonal $n \times n$ square matrices i.e. matrices O satisfying $O^t O = OO^t = 1$ where t denotes transpose.

Poincaré group: Inhomogeneous Lorentz group (Lorentz group plus space–time translations).

Pseudo-orthogonal groups: Groups whose elements are real square matrices satisfying the condition $O^t GO = OGO^t = G$ instead of the usual orthogonality condition, where G is a diagonal matrix with entries ± 1, i.e. G is an indefinite metric.

Rank l of compact simple Lie group: Maximal number of linearly independent commuting elements in Lie algebra (dimension of Cartan subalgebra).

Rigid group: Lie group whose parameters are not space–time dependent.

Symplectic group: $2n \times 2n$ matrix group whose elements G satisfy the condition $G^t J G = J$ where J is the skew-symmetric metric $J = \left(\begin{smallmatrix} 0 & 1 \\ -1 & 0 \end{smallmatrix}\right)$ and 1 is the $n \times n$ unit matrix. (J is the natural metric for the $2n$-dimensional phase-space of point mechanics.)

SYMMETRY BREAKING

Critical orbits: Orbits on which all the invariants are extremal.

Generic orbits: Orbits with minimal little group (dense in the orbit space).

Isotropy group (of a vector): Maximal closed subgroup which leaves the given vector invariant.

Little group: Isotropy group.

Maximal little group: One which is not contained in any other little group (in the same representation).

Minimal little group: One which contains no other little group (in the same representation).

Residual symmetry group: Isotropy group.

Soft symmetry breaking: Symmetry breaking by terms in the Lagrangian with dimension of mass or mass squared.

Spontaneous symmetry breaking: Symmetry breaking brought about by the fact that the Lagrangian (or Hamiltonian) is symmetric but the ground state (vacuum, minimum of the potential) is not.

Stability group (of potential): Isotropy group of vector for which the potential takes its minimum value.

Stability group (of vector): Isotropy group.

Index

Index